Digitising the Industry
Internet of Things Connecting the
Physical, Digital and Virtual Worlds

RIVER PUBLISHERS SERIES IN COMMUNICATIONS

Volume 49

Series Editors

ABBAS JAMALIPOUR
The University of Sydney
Australia

MARINA RUGGIERI
University of Rome Tor Vergata
Italy

HOMAYOUN NIKOOKAR
Delft University of Technology
The Netherlands

The "River Publishers Series in Communications" is a series of comprehensive academic and professional books which focus on communication and network systems. The series focuses on topics ranging from the theory and use of systems involving all terminals, computers, and information processors; wired and wireless networks; and network layouts, protocols, architectures, and implementations. Furthermore, developments toward new market demands in systems, products, and technologies such as personal communications services, multimedia systems, enterprise networks, and optical communications systems are also covered.

Books published in the series include research monographs, edited volumes, handbooks and textbooks. The books provide professionals, researchers, educators, and advanced students in the field with an invaluable insight into the latest research and developments.

Topics covered in the series include, but are by no means restricted to the following:

- Wireless Communications
- Networks
- Security
- Antennas & Propagation
- Microwaves
- Software Defined Radio

For a list of other books in this series, visit www.riverpublishers.com

Digitising the Industry
Internet of Things Connecting the
Physical, Digital and Virtual Worlds

Editors

Dr. Ovidiu Vermesan
SINTEF, Norway

Dr. Peter Friess
EU, Belgium

River Publishers

Published, sold and distributed by:
River Publishers
Alsbjergvej 10
9260 Gistrup
Denmark

River Publishers
Lange Geer 44
2611 PW Delft
The Netherlands

Tel.: +45369953197
www.riverpublishers.com

ISBN: 978-87-93379-81-7 (Hardback)
        978-87-93379-82-4 (Ebook)

©2016 River Publishers

Dedication

"To raise new questions, new possibilities, to regard old problems from a new angle, requires creative imagination and marks real advance in science."

— Albert Einstein

"When all think alike, then no one is thinking."

— Walter Lippman

Acknowledgement

The editors would like to thank the European Commission for their support in the planning and preparation of this book. The recommendations and opinions expressed in the book are those of the editors and contributors, and do not necessarily represent those of the European Commission.

Ovidiu Vermesan
Peter Friess

Contents

Preface

IoT – Key Enabling Pillar for Digitising Industry

The Internet of Things (IoT) is considered to be one of the enablers of the next industrial revolution. It is fuelled by the advancement of digital technologies, as well as dramatically changing how companies engage in business activities and people interact with their environment. The IoT's disruptive nature requires the assessment of the requirements for its future deployment across the digital value chain in various industries and many application areas.

The IoT is bridging the physical, digital, cyber and virtual worlds and requires sound information processing capabilities for the "digital shadows" of these real things. IoT applications are gradually moving from vertical, single-purpose solutions to multi-purpose and collaborative applications interacting across industry verticals, organizations and people, which represents one of the essential paradigms of the digital economy. Many of those applications still have to be identified, while involvement of end users in this innovation is crucial.

IoT technologies are key enablers of the Digital Single Market (DSM), which will have a potentially significant impact on the creation of jobs and growth, along with providing opportunities for IoT stakeholders in deploying and commercializing IoT technologies and applications within European and global markets.

The following chapters will provide insights into the state of the art for research and innovation regarding the IoT, while exposing you to the challenges and opportunities within future IoT ecosystems, which address IoT technology as well as applications developments and deployments for various domains (consumer/business/industrial/art).

Editors Biography

Dr. Ovidiu Vermesan holds a Ph.D. degree in microelectronics and a Master of International Business (MIB) degree. He is Chief Scientist at SINTEF Information and Communication Technology, Oslo. His research interests are in the area of mixed-signal embedded electronics and cognitive communication systems. Dr. Vermesan received SINTEFs 2003 award for research excellence for his work on the implementation of a biometric sensor system. He is currently working with projects addressing nanoelectronics, integrated sensor/actuator systems, communication, cyber-physical systems and the IoT, with applications in green mobility, energy, autonomous systems and smart cities. He has authored or co-authored over 75 technical articles and conference papers. He is actively involved in the activities of the Electronic Components and Systems for European Leadership (ECSEL) Joint Technology Initiative (JTI). He coordinated and managed various national, EU and other international projects related to integrated electronics. Dr. Vermesan is actively participating in national, H2020 EU and other international initiatives by coordinating and managing various projects. He is the coordinator and chair of the WG01 IoT European Research Cluster (IERC) of the Alliance for Internet of Things Innovation (AIOTI).

Dr. Peter Friess is a senior programme officer of DG CONNECT at the European Commission, taking care for more than ten years of the research and innovation policy for the Internet of Things. In his function he has shaped the on-going European research and innovation program on the Internet of Things and accompanied the European Commission's direct investment of over 150 Mill. Euro in this field. He also oversees the international cooperation on the Internet of Things, in particular with Asian countries. Furthermore, he integrates aspects of societal challenges, ethics and art linked to IoT. In previous engagements he was working as senior consultant for IBM, dealing with major automotive and utility companies in Germany and Europe. Prior to this engagement he worked as IT manager at Philips Semiconductors on business process optimisation in complex manufacturing, and before as researcher in

European and national research projects on advanced telecommunications and business process reorganisation. He is a graduated engineer in Aeronautics and Space technology from the University of Munich and holds a Ph.D. in Systems Engineering including self-organising systems from the University of Bremen. He also published a number of articles and co-edits a yearly book of the European Internet of Things Research Cluster.

List of Figures

List of Tables

1

Introduction

Peter Friess[1]

European Commission DG CONNECT, Belgium

The Internet of Things (IoT) has started to flourish excitingly. After having been in the expert corner for many years, new players and partners joined the field and contribute to manifest and extend the IoT. Business interest and novel ideas drive now the deployment. Today we do no longer question what IoT is or not, but more what solutions it can bring and what still needs to be done for a full blossom.

In the European policy context, the creation of a genuine Single Market encompasses the IoT as essential contribution. The European Commission gives indeed a strategic dimension to IoT for the Digital Single Market (DSM), not only in terms of regulatory challenges but also with regards to overcome interoperability issues and fragmented standards, probably one of the most dominant obstacles at the moment. The key objective remains a collaborative, responsible and fully functional IoT.

In the recently published IoT Staff Working Document[2], which has been elaborated based on extensive discussions with the IoT Community, we identify and describe 3 imperative pillars in order to advance IoT in Europe:

1. A single market for the IoT: IoT devices and services (thus including data) must be able to connect seamlessly and on a plug-and-play basis anywhere in the European Union (EU), and scale up without hindering from national borders;
2. A context of thriving IoT Ecosystems: new products and services in selected lead markets such as Industrial IoT, and the existence of

---

[1]The views expressed in this article are purely those of the author and may not, in any circumstances, be interpreted as stating an official position of the European Commission.

[2]https://ec.europa.eu/digital-single-market/en/news/staff-working-document-advancing-int ernet-things-europe

open platforms across vertical silos, helping developers' communities to innovate and not causing lock-in situations for users;

3. A human-centred IoT: European values must be translated in the design of IoT applications to empower citizens, and driven by high privacy and security standards and notably through a "Trusted IoT" label.

In order to work on these pillars, we launched seven innovation plus two coordination IoT Ecosystem projects in January 2016. They will be joined by a new round of IoT Large Scale Pilots already in January 2017, dealing with IoT scenarios in Assisted Living, Smart Agriculture, Wearables, Smart Cities and Connected Cars. The pilots will be complemented by accompanying measures on standardisation, security and privacy, creativity and art, further research on IoT platforms, and international calls.

Figure 1.1   Interactions within ecosystems.

In line with the ongoing cooperation with the IERC – the IoT European Research Cluster, the European Commission is equally committed to build upon the positive experience and to reinforce the cooperation with AIOTI – the Alliance for IoT Innovation for making Europe a leading IoT region. The Alliance has proven to be an important arena where frequently competing market actors can cooperate in order to improve interoperability issues of common interest and to contribute to the European IoT policy.

Besides the necessary emergence of IoT open platforms including neighbouring technologies, these are the subjects to work on for the next period: core standardisation, principles for appropriate design choices for technical and semantical interoperability, and increase of the trust level in IoT. As these questions do not allow to neglect the international dimension of IoT, we will be strategically interested in maintaining the cooperation with other leading IoT regions.

Looking ahead, we all are now establishing the first building blocks for a future hyper-connected society. Many new fascinating subjects such as smart objects, new interfaces for augmented realities and light forms of Artificial Intelligence will enter into the IoT applications and pave the way. Linked to it we will see many paradigm shifts, from a stronger consideration of environmental aspects and towards the transformation of competition to co-creation.

IoT is the future.

2

IoT Ecosystems Implementing Smart Technologies to Drive Innovation for Future Growth and Development

Peter Friess[1,2] and Rolf Riemenschneider[1,2]

[1]European Commission DG CONNECT, Belgium

"What is it good for, if not for Human Mankind?"

2.1 Introduction

In the early 1990s, James F. Moore was at the origin of the concept of business ecosystems [1], now becoming an interesting approach for the design of Internet of Things (IoT) evolution and deployment.

Moore defined "business ecosystem" as "an economic community supported by a foundation of interacting organizations and individuals – the organisms of the business world. The economic community produces goods and services of value to customers, who are themselves members of the ecosystem. The member organisms also include suppliers, lead producers, competitors, and other stakeholders. Over time, they coevolve their capabilities and roles, and tend to align themselves with the directions set by one or more central companies. Those companies holding leadership roles may change over time, but the function of ecosystem leader is valued by the community because it enables members to move toward shared visions to align their investments, and to find mutually supportive roles".

Given the current state of IoT evolution, and the complexity of IoT systems and actors involved, applying the concept of ecosystem is highly promising; in particular for two reasons:

---

[2]The views expressed in this article are purely those of the author and may not, in any circumstances, be interpreted as stating an official position of the European Commission.

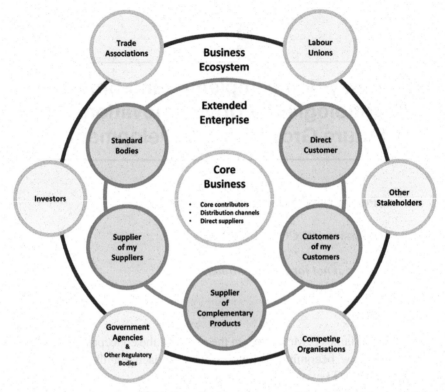

Figure 2.1   Business ecosystem [1].

- the nature of IoT itself prompts towards new ways of conceiving ICT systems, and changing the understanding of business and interaction processes and,
- a multitude of service providers involved whose role can change over time.

The European Commission has thus decided to apply this concept for its current IoT research and innovation policy. This concept is often similarly specified as either IoT Ecosystem, IoT Innovation Ecosystem or IoT Business Ecosystem; for reasons of simplicity we will talk here only about IoT Ecosystems (used in plural as there will be one or more IoT Ecosystems). Although this concept is certainly of universal nature, we will focus on IoT Ecosystems with a European center of gravity – less in the sense of a limitation but more as an operational vector of European values.

As the concept of IoT includes both a vertical and a horizontal dimension, a key feature of an IoT Ecosystem will therefore be the dynamic interaction between the providers and users of horizontal IoT platforms and applications and the providers and users of vertical solutions/domain-specific environments. Evolution of the IoT will also bring new devices to the market, around which IoT Ecosystems will take shape, and the IoT will act as an essential driver for innovation and competitiveness. More jobs are expected to be created, driven by the need for developers to work on applications and interfaces. While today around 300,000 developers worldwide contribute to the IoT, a new report by VisionMobile [2] projects 4.5 million developers by 2020, reflecting a 57% compound annual growth rate and a massive opportunity. As a consequence, the need is arising for well-educated employees in terms of education and training in the EU, having the necessary digital and interaction skills.

2.2 Support for IoT Ecosystem Creation

Whereas it might very be tempting to apply observations from natural and biological ecosystems onto human social and economic systems, several factors indicate that a one to one translation is not directly possible – humans through their brain evolution have a different understanding of existence that other species in natural systems whose primary objective is survival.

Figure 2.2 Self-organising systems [9].

Moreover, the necessary system thinking for ecosystems is a radical change compared with a former system thinking from the last century, where the control paradigm and a more technical understanding of systems were principally dominating. In modern system theory, this understanding has been replaced by an evolutionary approach and the vanishing of the idea of a central controller. The present system thinking is based on self-organisation, self-reference, co-evolution rather than opposition, and a more dynamic understanding of time. Following this approach, the influence of IoT Ecosystems is possible through the setting of boundary conditions; however, any direct influence should be avoided as the ecosystem will resist or ignore this input [3, 4].

In order to provide suitable boundary conditions for future IoT Ecosystems, the European Commission, since 2014, has launched the following actions:

- Stimulation of IoT Community building through the IERC – IoT European Research Cluster, in particular extending the outreach of ongoing projects through platform creation and international cooperation.
- Preparation for the creation of an industry-driven Alliance for IoT Innovation which was established in 2015 and counts today 500 members and 13 dedicated workgroups, based on the condition that members should possess a strong foothold in Europe.
- Launch of a 51 MEUR call for proposals on large projects for IoT Ecosystems in 2014 as part of the innovation programme Horizon 2020, linked to platforms for connected and smart objects. This call included from a systemic perspective a mix of Innovation actions and complementary Support actions for overcoming the fragmentation of vertically-oriented closed systems, architectures and application areas. Up to 10 MEUR were targeted to SMEs and start-ups working with these platforms.
- A series of community building events, gathering more than 300 people for discussing the call for IoT Ecosystems and providing a platform for federation and a learning space.
- Preparation of an IoT Large Scale Pilot innovation programme with a corresponding funding of around 100 MEUR for 2016–17, addressing Smart Agriculture, Assisted Living, Wearables, Smart Cities and Connected semi-autonomous cars. This will be coupled with a dedicated subsequent call on future IoT architectures, concepts, methods and tools for open IoT platforms advanced concepts for end-to-end IoT security and privacy (35 MEUR).

- Fuelling the IoT community with input from leading and large IoT-deploying regions such as Japan, Korea, and Brazil through joint calls.
- Opening of the IoT innovation area to new players from the Cloud, Big Data, Semi-autonomous systems and 5G domains, and to creativity and art – makers, innovation hubs, geeks and artists (not to forget the STARTS [5] initiative).
- Creation of an IoT Focus Area for improving coordination across Units, Directorates and Directorate-Generals of the European Commission and for providing a more centralised entry point to IoT.

These activities are complementary with various IoT initiatives in European Member States and should not be perceived in isolation to further European initiatives such the Digitising European Industry strategy.

2.3 Spurring Innovation in Lead Markets

With industry players all battling to "own" customers and their data, the IoT market looks chaotic and fragmented.

Tangible business opportunities for IoT technologies can be found across all "smart" environments identified by various experts. By combining estimated market size and growth potentials, some of these environments have emerged as offering the most realistic opportunities between now and the coming five years.

The challenge is to foster the deployment of IoT solutions in Europe through integration of advanced IoT technologies across the value chain, demonstration of multiple IoT applications at scale and in a usage context as close as possible to operational conditions. Compared to existing solutions, the roadblocks to overcome include i) the integration and further research and development, where needed, of the most advanced technologies across the value chain (components, devices, networks, middleware, service platforms, application functions) and their operation at large scale to respond to real needs of end-users (public authorities, citizens and business), based on underlying open technologies and architectures that may be reused across multiple use cases and enable interoperability across those; ii) the validation of user acceptability by addressing, in particular, issues of trust, attention, security and privacy through pre-defined privacy and security impact assessments, liability, coverage of user needs in the specific real-life scenarios of the pilot,

Figure 2.3   IoT industry's fragmentation [10].

iii) the validation of the related business models to guarantee the sustainability of the approach beyond the project.

The most prominent "smart" environments, already producing a number of use cases, are the following:

- Smart Homes will offer business opportunities in home security, energy applications and household appliances.
- Personal Wellness applications and wearable devices for both generic and health-specific purposes are a big opportunity in the area of Smart Health. They will be accompanied by remote health monitoring.
- In Smart Manufacturing, operations and asset management already represent a fertile ground for IoT solutions and applications.
- Smart Cities are equipped with sensors, actuators and other appliances providing information that, properly valorised, will improve the living conditions of their inhabitants.
- Smart Mobility will require new mobile ecosystems based on trust, security and convenience in order to ensure the security and convenience of consumer-centric transactions and services.
- For Smart Energy, smart meters and smart grids are powered by IoT and can optimise energy consumption, whereas IoT solutions and services can help change behaviour and consumption patterns.

- In Smart Farming data gathering, data processing, data analysis and automation technologies jointly orchestrated allow for improved operation and management of a farm and further down the value chain.
- Earth and ocean observation systems and the future blue economy where IoT can help maximise the use of oceans' potential, in terms of fishing, marine platforms and aquaculture notably.
- For the Circular Economy, IoT can facilitate the transition to new business models where all actors of the value chains are closely interconnected and use collaborative platforms to share data on resource flows, and end-users are empowered in their consumption patterns.

The Alliance for IoT Innovation – AIOTI has established a number of working groups in the areas that it considers more mature for IoT innovation and where a greater potential for cross-cutting business models is looming ahead [6]. In addition, the support of creativity-based innovation is pivotal, adding the force of cultural and creative industries to foster smart, sustainable and inclusive IoT services and products.

As an outcome of extended consultations and studies, the European Commission has decided to finance IoT pilot projects with a larger scope and a potential for changing the perception and acceptance in the following fields:

- Pilot 1: Smart living environments for ageing well (EU contr. up to 20 MEUR)
- Pilot 2: Smart Farming and Food Security (EU contr. up to 30 MEUR)
- Pilot 3: Wearables for smart ecosystems (EU contr. up to 15 MEUR)
- Pilot 4: Reference zones in EU cities (Smart Cities) (EU contr. up to 15 MEUR)
- Pilot 5: Autonomous vehicles in a connected environment (EU contr. up to 20 MEUR)

These pilots are complemented by two categories of support actions:

- Co-ordination of and support to the pilots through mapping of architecture approaches; interoperability and standards approaches at technical/semantic levels; requirements for legal accompanying measures; common methodologies for design, testing and validation; federation of pilot activities and transfer; exploitation of security and privacy mechanisms, international cooperation and exploitation of combination of ICT and Art.
- Consideration of responsible innovation and societal aspects, also through involvement of experts outside the traditional field of IoT.

It is expected that these IoT Large Scale Pilot projects will enter into action as of January 2017, complementing the already active IoT Ecosystem projects from the previous call for proposals, now brought under the common umbrella IoT European Platform Initiative – IoT-EPI [7]. Conceptually the future IoT Large Scale Pilot projects are a variant in terms of IoT Ecosystem building and target in particular innovation integration and the overcoming of acceptance, adoption and legislative barriers against wide-ranging IoT deployment.

2.4 Outlook

Looking forward, we can contemplate that the current and upcoming IoT activities, when properly set up, will contribute a lot to the birth and evolution of IoT Ecosystems in Europe.

The recent EC Digital Single Market (DSM) technologies and public services modernisation package provides a set of coherent policy measures aiming at the digital transformations of our industries and at maximising their impact on economic growth. The actions for IoT are listed in the communication "Digitising European Industry – Reaping the full benefits of a Digital Single Market [8], the communication "Priorities for ICT Standardisation for the Digital Single Market", and under the free flow of data initiative of the DSM Strategy.

Fostering an interoperable environment for IoT Ecosystems and the development of missing interoperability standards will be pivotal. Exploration of options and guiding principles, including developing standards for trust, privacy and end-to-end security, e.g. through a 'trusted IoT label', are equally high on the policy agenda.

With regards the Horizon work programme 2018–20 for IoT, it is expected to support IoT Large Scale Pilot initiatives of societal and industrial relevance and to facilitate use cases crossing existing IoT pilots and implementations, both in Europe and with international partners. In addition, the existing IoT Focus Area might also encompass more aspects of Cloud technologies, Big Data analysis, autonomous behaviour, interface technologies and art.

Bibliography

[1] J. F. Moore, *The Death of Competition: Leadership and Strategy in the Age of Business Ecosystems*, HarperBusiness, 1996.

[2] VisionMobile, *IoT Megatrends 2015*, online at http://www.visionmobile.com/product/iot-megatrends-2015/

[3] H. von Foerster, *Understanding understanding*, Springer, 2002.

[4] J. Rifkin, *The Zero Marginal Cost Society*, Palgrave McMillan, 2014.

[5] STARTS, online at https://ec.europa.eu/digital-single-market/en/ict-art-starts-platform

[6] AIOTI, online at www.aioti.eu

[7] IoT-EPI, online at http://iot-epi.eu/

[8] Digitising European Industry, online at https://ec.europa.eu/digital-single-market/en/digitising-european-industry

[9] P. Miller, *The Smart Swarm*, The Penguin Group, 2010.

[10] J. Yoshida, *Google, Silicon Labs mesh for ZigBee-like protocol*, online at http://www.analog-eetimes.com/news/google-silicon-labs-mesh-zigbee-protocol

3

IoT Digital Value Chain Connecting Research, Innovation and Deployment

Ovidiu Vermesan[1], Peter Friess[2], Patrick Guillemin[3], Martin Serrano[4], Mustapha Bouraoui[5], Luis Pérez Freire[6], Thomas Kallstenius[7], Kit Lam[8], Markus Eisenhauer[9], Klaus Moessner[10], Maurizio Spirito[11], Elias Z. Tragos[12], Harald Sundmaeker[13], Pedro Malo[14] and Arthur van der Wees[15]

[1]SINTEF, Norway
[2]European Commission, Belgium
[3]ETSI, France
[4]Digital Enterprise Research Institute, Galway, Ireland
[5]STMicroelectronics, France
[6]GRADIANT, Spain
[7]iMinds vzw, Belgium
[8]SAMSUNG Electronics Research and Development Institute, UK
[9]Fraunhofer FIT, Germany
[10]University of Surrey, UK
[11]ISMB, Italy
[12]FORTH, Greece
[13]ATB Institute for Applied Systems Technology Bremen, Germany
[14]FCT NOVA and UNINOVA, Portugal
[15]Arthur's Legal B.V., The Netherlands

"Productivity isn't about how busy or efficient you are – it's about how much you accomplish." Chris Bailey

3.1 Internet of Things Vision

Internet of Things (IoT) is considered one of the next industrial revolution enablers, which is fuelled by the advancement of digital technologies. IoT is dramatically changing how companies engage in business activities, and

15

how people will interact with their environment. Its disruptive nature requires the assessment of the requirements for the future deployment across the digital value chain in various industries and in many application areas.

IoT is a concept and a paradigm with different visions, and multidisciplinary activities. IoT considers pervasive presence in the environment of a variety of things, which through wireless and wired connections and unique addressing schemes are able to interact with each other and cooperate with other things to create new applications/services and reach common goals. In the last few years IoT has evolved from being simply a concept built around communication protocols and devices to a multidisciplinary domain where devices, Internet technology, and people (via data and semantics) converge to create a complete ecosystem for business innovation, reusability, interoperability, that includes solving the security, privacy and trust implications.

The IoT is the network of physical objects that contain embedded technology to communicate and sense or interact with their internal states or the external environment. The confluence of efficient wireless protocols, improved sensors, cheaper processors, and a bevy of startups and established companies developing the necessary management and application software, has finally made the concept of the IoT mainstream. The IoT makes use of synergies that are generated by the convergence of Consumer, Business and Industrial Internet customer, Business and Industrial Internet. The convergence creates the open, global network connecting people, data, and things. This convergence leverages the cloud to connect intelligent things that sense

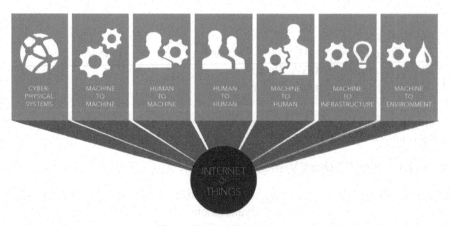

Figure 3.1   IoT integration.

and transmit a broad array of data, helping creating services that would not be obvious without this level of connectivity and analytical intelligence. The dynamics surrounding emerging IoT applications are very complex and issues such as enablement, network connectivity, systems integration, value-added services, and other management functions are all needs that generally must be addressed when the end-users seek to connect intelligent edge devices into complex IoT applications [59].

In this context, the research and development challenges to create a smart world are enormous. IoT ecosystems offer solutions comprising of large heterogeneous systems of systems beyond an IoT platform and solve important technical challenges in the different industrial verticals and across verticals.

IoT's disruptive nature requires the assessment of the requirements for the future deployment across the digital value chain in various industries and in many application areas considering even better exchange of data, the use of standardized interfaces, interoperability, security, privacy, safety, trust that will generate transparency, and more integration in all areas of the Internet (consumer/business/industrial).

IoT will generate even more data that needs to be processed and analysed, and the IoT applications will require new business models and product-service combinations to address and tackle the challenges in the Digital Single Market (DSM).

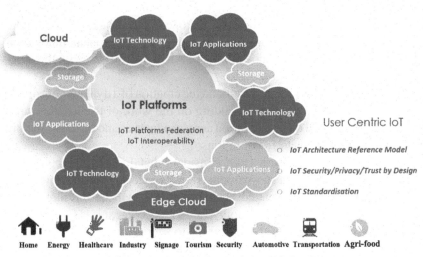

Figure 3.2   IoT platforms interaction and federation.

The use of platforms is being driven by transformative technologies such as cloud, things, and mobile. The IoT and services makes it possible to create networks incorporating the entire manufacturing process that convert factories into a smart environment. The cloud enables a global infrastructure to generate new services, allowing anyone to create content and applications for global users. Networks of things connect things globally and maintain their identity online. Mobile networks allow connection to this global infrastructure anytime, anywhere. The result is a globally accessible network of things, users, and consumers, who are available to create businesses, contribute content, generate and purchase new services.

Platforms also rely on the power of network effects, as they allow more things, they become more valuable to the other things and to users that make use of the services generated. The success of a platform strategy for IoT can be determined by connection, attractiveness and knowledge/information/data flow.

In this context, the Alliance for Internet of Things Innovation (AIOTI), was initiated following the European and global IoT technology and market developments.

The aim of AIOTI is to create and master sustainable innovative European IoT ecosystems in the global context to address the challenges of IoT technology and applications deployment including standardisation, interoperability and policy issues, in order to accelerate sustainable economic development and growth in the new emerging European and global digital markets. The AIOTI is connecting/integrating technologies and applications across the digital value chain and has strong links with the other European initiatives (Private Public Partnerships – PPPs, Joint Technology Initiatives – JTIs, European Innovation Partnerships – EIPs, etc.). The positioning of AIOTI in relation with the other initiatives is presented in Figure 3.3.

The members of AIOTI jointly work on the creation of a dynamic European IoT ecosystem. This ecosystem is building on the work of the IoT Research Cluster (IERC) and spill over innovation across industries and business sectors of IoT transforming ideas to IoT solutions.

The European Commission (EC) considers that IoT will be pivotal in enabling the DSM, through new products and services. The IoT, big data, cloud computing and their related business models will be the three most important drivers of the digital economy, and in this context it is fundamental for a fully functional single market in Europe to address aspects of ownership, access, privacy and data flow – the new production factor.

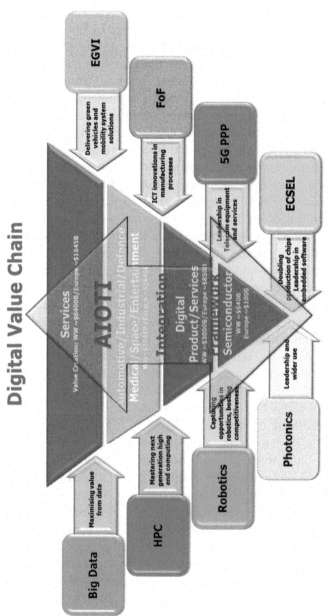

Figure 3.3 AIOTI integration framework.

3.1.1 IoT Common Definition

The IoT is a key enabling technology for digital businesses and one of the main drivers that is contributing to transform the Internet. IoT technologies are deployed in different sectors, from agricultural in rural areas, health and wellness to smart home and smart-X applications in cities.

The IoT is bridging the virtual, digital, physical worlds and mobile networks need to scale to match the demands of billions of things, while the processing capabilities require addressing the information provided by the "digital shadow" of these real things. This need focusing on the developments in the virtual world and the physical world for solving the challenges of IoT applications. In the virtual world, network virtualization, software-defined hardware/networks, device management platforms, edge computing and data processing/analytics are developing fast and urgency to be endeavoured as enabling technologies for IoT. Connecting the virtual, digital, physical worlds generates knowledge through IoT applications and platforms, while addressing security, privacy and trust issues across these dimensions.

Smart IoT applications modify the way people interact with the intelligent spaces (called also cyber-spaces), from how remotely control appliances at home to how the care for patients or elderly persons are performed. The massive deployment of IoT devices represents a tremendous economic impact and at the same time offers multiple opportunities. The IoT's potential is underexploited, the physical and intelligent are largely disconnected, requiring a lot of manual effort to find, integrate, and use information in a meaningful way. IoT and its advances in intelligent spaces advances can be categorised along with the key technologies at the core of the Internet.

Intelligent spaces are created and enriched by IoT and they are environments in which ICTs, sensor and actuator systems become embedded into physical objects, infrastructures, and the places embedded of technology that facilitate physical-human-cyber communication named intelligent surroundings or cyber places in which people live, (e.g. smart cities, industrial/manufacturing plants, homes and buildings, automotive and entertainment). The goal is to enable computers and smart edge devices to take part in activities never previously involved and people to interact with computers and these devices at the edge more naturally i.e. gesture, voice, movement, and context, etc. The IoT developments in the various sectors has created IoT ecosystems that are focusing on Internet of X technologies and applications that address the specific needs of the respective sector with the goal to be interoperable across various other sectors as presented in Figure 3.5.

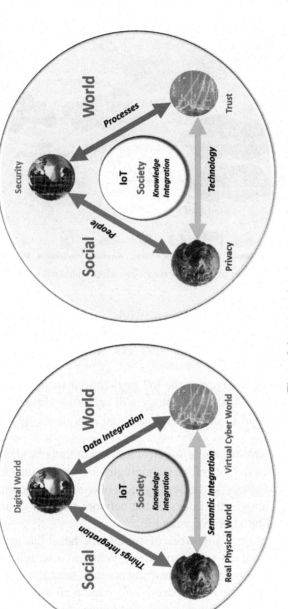

Figure 3.4 Integration.

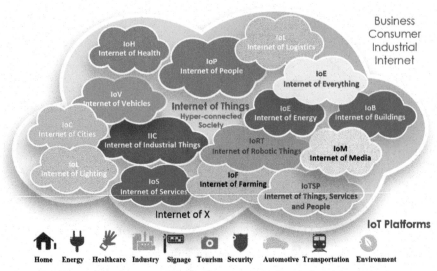

Figure 3.5   Internet of X developments in various industrial sectors.

The traditional distinction between network and device is starting to blur as the functionalities of the two become indistinguishable. Shifting the focus from the IoT network to the devices costs less, scales more gracefully, and leads to immediate revenues.

As a result of this convergence, the IoT applications require that classical industries are adapting and the technology will create opportunities for new industries to emerge and to deliver enriched and new user experiences and services.

In addition, to be able to handle the sheer number of things and objects that will be connected in the IoT, cognitive technologies and contextual intelligence are crucial. This also applies for the development of context aware applications that need to be reaching to the edges of the network through smart devices that are incorporated into our everyday life.

The Internet is not only a network of computers, but it has evolved into a network of devices of all types and sizes, vehicles, smartphones, home appliances, toys, cameras, medical instruments and industrial systems, all connected, all communicating and sharing information all the time.

The IoT has different meanings at different levels of abstractions through the value chain, from lower level semiconductor through the service providers. IoT is a paradigm with different visions, and involving multidisciplinary activities.

The IoT as a "global concept" requires a common high-level definition. Considering the wide background and required technologies, from sensing device, communication subsystem, data aggregation and pre-processing to the object instantiation and finally service provision, generating an unambiguous definition of the "IoT" is non-trivial.

The IERC is actively involved in ITU-T Study Group 13, which leads the work of the International Telecommunications Union (ITU) on standards for next generation networks (NGN) and future networks and has been part of the team which has formulated the following definition [42]: "Internet of things (IoT): A global infrastructure for the information society, enabling advanced services by interconnecting (physical and virtual) things based on existing and evolving interoperable information and communication technologies. NOTE 1 – Through the exploitation of identification, data capture, processing and communication capabilities, the IoT makes full use of things to offer services to all kinds of applications, whilst ensuring that security and privacy requirements are fulfilled. NOTE 2 – From a broader perspective, the IoT can be perceived as a vision with technological and societal implications.

The IERC definition [45] states that IoT is "A dynamic global network infrastructure with self-configuring capabilities based on standard and interoperable communication protocols where physical and virtual "things" have identities, physical attributes, and virtual personalities and use intelligent interfaces, and are seamlessly integrated into the information network".

3.1.2 Artificial Intelligence and Cognitive IoT

IoT applications are generating data collected from various domains and industrial sectors. The data generated provides insights from the environments and applications that generated it. Artificial Intelligence (AI) techniques provide the framework and tools to go beyond analytics of real time monitoring and automation use cases for IoT and move to IoT platforms that use concepts from artificial intelligence and apply them to specific IoT use cases to provide smarter decision-making. AI-enabled IoT applications add a new layer of functionality and access, creating the next generation of smart homes/buildings, smart vehicles, and smart manufacturing by providing intelligent automation, predictive analytics and proactive intervention.

In the IoT context, AI will support companies in finding the smart data and analyse the trends and patterns for better decision-making based on defined set of rules.

The AI techniques will enable cognitive systems to be integrated with IoT applications creating optimized solutions for each individual applications. Cognitive IoT technologies will allow embedding intelligence into systems and processes, allowing businesses to increase efficiency, find new business opportunities, and to anticipate risks and threats so they can better address them. The IoT applications will gather and integrate data from many types of sensors and other sources, reason over data, and learn from the interactions, while creating communities of devices that share information. The information collected can be interpreted and managed by people, IoT applications or IoT platforms using cognitive systems in order to generate new and better services and use cases.

The data generated by edge devices combined with the unstructured data available from sources ranging from news Web sites and social networks can be combined using cognitive IoT capabilities at the edge or at the cloud level.

The use on artificial intelligence, swarm intelligence and cognitive technologies together with deep learning techniques for optimising the IoT services provided by IoT applications in smart environments and collaboration spaces will create solutions capable of transforming industries and professions.

3.1.3 IoT of Robotic Things

IoT, artificial intelligence, robotics, machine learning, swarm technologies are the technologies that will provide the next phase of development of IoT applications. Robotics provide the programmed machines designed to be involved in labour intensive and repetitive work, while deep machine learning is the science of allowing/empowering machines to function using learning algorithms instead of programing. The combination of these disciplines opens the developments of autonomous systems combining robotics and machine learning for designing robotic systems to autonomous. Machine learning is part of advance state of intelligence using statistical pattern recognition, parametric/non-parametric algorithms, neural networks, recommender systems, swarm technologies etc. in order to perform autonomous tasks. The industrial IoT is a subset of the IoT, where edge devices, processing units and networks interact with their environments to generate data to improve processes.

The IoT, the technologies, architectures, and services that allow massive numbers of sensor enabled, uniquely addressable "things" to communicate with each other and transfer data over pervasive networks using Internet protocols, is expected to be the next great technological innovation and business opportunity. Many IoT initiatives are focused on using connected devices with edge devices to manage, monitor and optimize systems and their processes. Advanced and transformational aspects of ubiquitous connectivity and communication include intelligent devices that monitor events, fuse sensor data from a variety of sources, use local and distributed "intelligence" to determine a best course of action, and then act to control or manipulate objects in the physical world, and in some cases while physically moving through that world. The concept called Internet of Robotic Things (IoRT), addresses the many ways IoT technologies and robotic "devices" intersect to provide advanced robotic capabilities, along with novel applications, and by extension, new business, and investment opportunities [17].

The combination of advanced sensing, communication, local and distributed processing, and actuation take the original vision for the IoT to

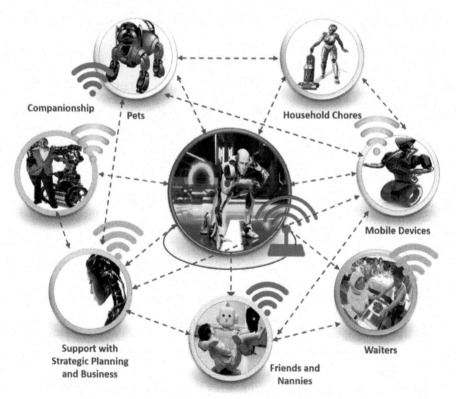

Figure 3.6 Internet of Robotic Things (IoRT) pervasive technology.

a wholly different level, and one that opens up completely new classes of opportunities for IoT and robotics solution providers, as well as users of their products. The concept allows to:

- Define and describe the characteristics of robotics technologies that distinguish them as a separate, unique class of IoT objects, and one that differs considerably from the common understanding of IoT edge nodes as simple, passive devices.
- Reveal how the key features of robotics technology, namely movement, mobility, manipulation, intelligence and autonomy, are enhanced by the IoT paradigm, and how, in turn, the IoT is augmented by robotic "objects" as "intelligent" edge devices.
- Illustrate how IoT and robotics technologies combine to provide for Ambient Sensing, Ambient Intelligence and Ambient Localization, which can be utilised by new classes of applications to deliver value.

IoT, cognitive computing and artificial intelligence are very important to the strategies for digital value chain integration addressing the implementation of IoT applications in various smart environments.

3.2 IoT Strategic Research and Innovation Directions

The IERC is bringing together EU funded projects with the aim of defining a common vision of IoT technology and addressing European research challenges. The rationale is to target at the large potential for IoT-based capabilities and promote the use of the results from the existing projects to encourage convergence of ongoing work to tackle the most important deployment issues and the transfer of research and knowledge to products and services and apply these in real IoT applications. The vison is illustrated in Figure 3.7 [59].

IoT is a new revolution of the Internet. Things make themselves recognizable and they obtain intelligence thanks to the fact that they can communicate information about themselves and they can access information that has been aggregated by other things.

The technological trend is a move from systems where there are multiple users/people per device, people in control loop of the system, and the system providing the ability for people to interact with people. The IoT brings a new paradigm where there are multiple devices per user; the devices are things that are connected and communicating with other things. The interaction will be with a heterogeneous continuum of users, things and real physical events

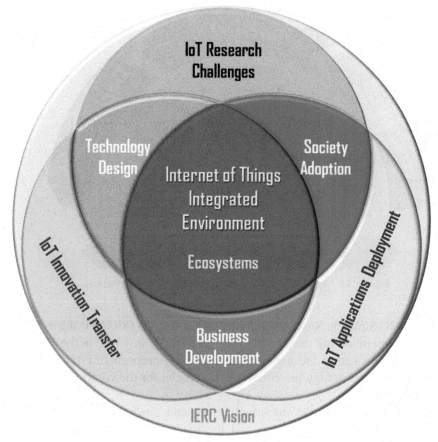

Figure 3.7 IERC Vison for IoT integrated environment and ecosystems.

(e.g., move left/right/up/down, change humidity/temperature/light/sound, etc.) and the Internet is the common convergence connectivity capability, replacing the previous independent systems.

The objectives of IERC is to provide the research and innovation trends, presenting the state of the art in terms of IoT technology and societal analysis in order to apply the develop to the IoT funded projects and further into the market applications and in the EU policies. The final goal is to test and develop innovative and interoperable IoT solutions in areas of industrial and public interest. The IERC objectives are addressed as an IoT continuum of research, innovation, development, deployment and adoption as presented in Figure 3.8 [59].

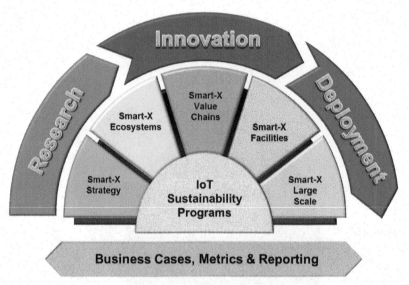

Figure 3.8   IoT continuum: research, innovation, deployment.

The IERC Strategic Research and Innovation Agenda (SRIA) is the result of a discussion involving the projects and stakeholders involved in the IERC activities, which gather the major players of the European ICT landscape addressing IoT technology priorities that are crucial for the competitiveness of European industry.

IERC SRIA covers the important issues and challenges for the IoT technology. It provides the vision and the roadmap for coordinating and rationalizing current and future research and development efforts in this field, by addressing the different enabling technologies covered by the IoT concept and paradigm.

The future IoT developments will address highly distributed IoT applications involving a high degree of distribution, and processing at the edge of the network by using platforms that that provide compute, storage, and networking services between edge devices and computing data centres. These platforms will support emerging IoT applications that demand real-time latency (i.e. mobility/transport, industrial automation, safety critical wireless sensor networks, etc.). These developments will bring new challenges as presented in Figure 3.9 [59].

The IoT value will come from the combination of edge computing and data centre computing considering the optimal business model, the right location, right timing, and efficient use of available network resources and bandwidth.

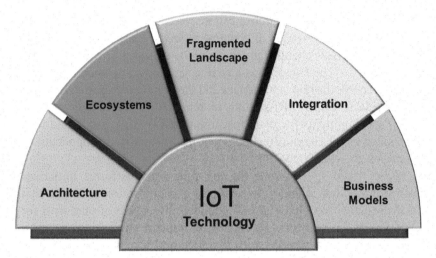

IoT Technology and Applications Research and Development

Figure 3.9   IoT future challenges.

The IoT architecture, like the Internet, will grow in evolutionary fashion from a variety of separate contributions and there are many current efforts regarding architecture models under development. The challenges for the IoT architecture are the complexity and cooperative work for developing, adopting and maintaining an effective cross-industry technology reference architecture that will allow for true interoperability and ease of deployment.

The IERC will work for providing the framework for the convergence of the IoT architecture approaches considering the vertical definition of the architectural layers end-to-end security and horizontal interoperability. IoT technology is deployed globally, and supporting the activities for common unified reference architecture would increase the coherence between various IoT platforms. A common architectural approach will require focusing on the reference model, specifications, requirements, features and functionality. In particular, this issue would be important in preparation of the future IoT LSPs, although time schedule might be difficult to synchronize.

The IERC SRIA is developed with the support of a European-led community of interrelated projects and their stakeholders, dedicated to the innovation, creation, development and use of the IoT technology.

Since the release of the first version of the IERC SRIA, we have witnessed active research on several IoT topics. On the one hand this research filled several of the gaps originally identified in the SRIA, whilst on the other it created new challenges and research questions. Recent advances in areas such as cloud computing, cyber-physical systems, robotics, autonomic computing, and social networks have changed the scope of the Internet of Thing's convergence even more so. The Cluster has a goal to provide an updated document each year that records the relevant changes and illustrates emerging challenges. The updated release of this SRIA builds incrementally on previous versions [45, 46, 73] and highlights the main research topics that are associated with the development of IoT enabling technologies, infrastructures and applications with an outlook towards 2020 [51].

The research activities include the IoT European Platforms Initiatives (IoT-EPI) program that includes the research and innovation consortia that are working together to deliver an IoT extended into a web of platforms for connected devices and objects. The platforms support smart environments, businesses, services and persons with dynamic and adaptive configuration capabilities. The goal is to overcome the fragmentation of vertically-oriented closed systems, architectures and application areas and move towards open systems and platforms that support multiple applications. IoT-EPI is funded by the European Commission (EC) with EUR 50 million over three years.

The projects involved in the programs are listed in the Figure 3.10. The projects are part of the IERC and are cooperating to define the research and innovation mechanisms and identify opportunities for collaboration in IoT ecosystems to maximise the opportunities for common approaches to platform development, interoperability and information sharing. The common activities are organised under six task forces (Figure 3.11) that are conceived and developed under the IoT-EPI program.

The task forces are complementary to the IERC activity chains. The activity chains are created to favour close cooperation between the IoT Cluster projects, the IoT-EPI programme and the AIOTI working groups to form an arena for exchange of ideas and open dialog on important research challenges. The activity chains are defined as work streams that group together partners or specific participants from partners around well-defined technical activities that will result into at least one output or delivery that will be used in addressing the IERC objectives.

The research and innovation items addressed and discussed in the task forces of the IoT-EPI program, the IERC activity chains and the AIOTI working groups for the basis of the IERC SRIA that addresses the roadmap of IoT technologies and applications in line with the major economic and societal challenges underlined in the EU 2020 Digital Agenda [52].

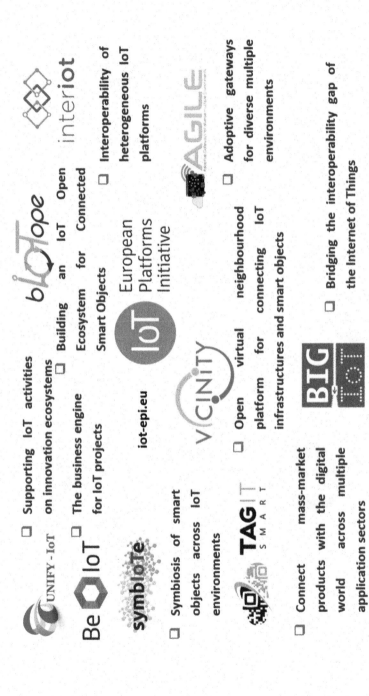

Figure 3.10 IoT-EPI program projects.

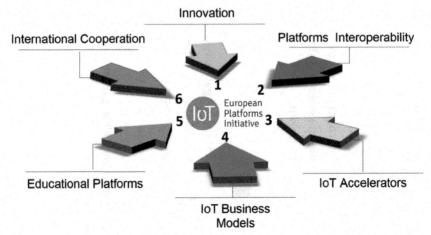

Figure 3.11   IoT-EPI task forces.

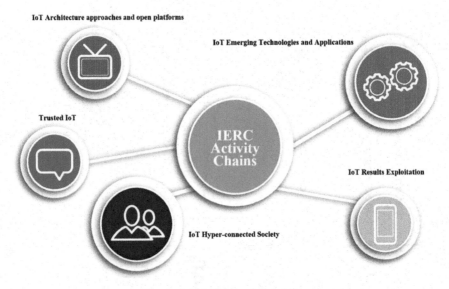

Figure 3.12   IERC activity chains.

The IERC SRIA is developed incrementally based on its previous versions and focus on the new challenges being identified in the last period.

The updated release of the SRIA is highlighting the main research topics that are associated with the development of IoT infrastructures and applications, with an outlook towards 2020 [51].

The timeline of the IERC IoT SRIA covers the current decade with respect to research and the following years with respect to implementation of the research results. As the Internet and its current key applications show, it is anticipated that unexpected trends will emerge leading to unforeseen new development paths.

The IERC has involved experts working in industry, research and academia to provide their vision on IoT research challenges, enabling technologies and the key applications, which are expected to arise from the current vision of the IoT.

The multidisciplinary nature of IoT technologies and applications is reflected in the IoT digital holistic view adapted from [32].

IoT demands an extensive range of new technologies and skills that many organizations have yet to master and creates challenges for organizations exploiting the IoT. The technologies and principles of IoT will have a very broad impact on organizations, affecting business strategy, risk management and a wide range of technical areas such as architecture and network design. The top 10 IoT technologies for 2017 and 2018 as presented by Gartner [21] are:

- IoT Security – due to hardware and software advances IoT security is a fast-evolving area through 2021 and the skills shortage today will only accelerate. Enterprises need to begin investing today in developing this expertise in-house and begin recruitment efforts. Many security problems

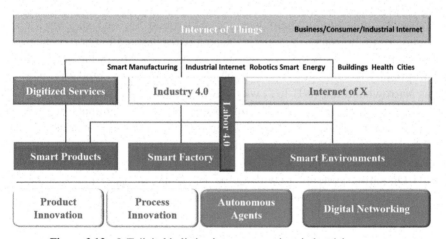

Figure 3.13 IoT digital holistic view across various industrial segments.

today are the result of poor specification, design, implementation and lack of knowledge/training. It is expected that the companies adopting IoT are investing in these areas.

- IoT Analytics – that require new algorithms, architectures, data structures and approaches to machine learning if organizations are going to get the full value of the data captured, and knowledge created. Distributed analytics architectures to capitalize on pervasive, secure IoT network architectures will evolve into become knowledge sharing networks.
- IoT Device Management – Significant innovation will result from the challenges of enabling technologies that are context, location, and state-aware while at the same time consistent with data and knowledge taxonomies. IoT Device Management will probably break the boundaries of traditional data management and create data structures capable of learning and flexing to unique inbound data requirements over time.
- Low-Power, Short-Range IoT Networks – Low-power, short-range networks will dominate wireless IoT connectivity through 2025, far outnumbering connections using wide-area IoT networks.
- Low-Power, Wide-Area Networks – traditional cellular networks cannot deliver a proper combination of technical features and operational cost for those IoT applications that need wide-area coverage combined with relatively low bandwidth, good battery life, low hardware and operating cost, and high connection density. Wide-area IoT networks aim is to deliver data rates from hundreds of bits per second (bps) to tens of kilobits per second (kbps) with nationwide coverage, a battery life of up to 10 years, an endpoint hardware cost of around $5, and support for hundreds of thousands of devices connected to a base station or its equivalent. The first low-power wide-area networks (LPWANs) were based on proprietary technologies, but in the long term, emerging standards such as Narrowband IoT (NB-IoT) will likely dominate this space.
- IoT Processors – low-end 8-bit microcontrollers will dominate the IoT through 2019 and shipments of 32-bit microcontrollers will overtake the 8-bit devices by 2020. The report does not mention the 16-bit processors ever attaining critical mass in IoT applications.
- IoT Operating Systems – a wide range of IoT-specific operating systems with minimal and small footprint will gain momentum in IoT through 2020 as traditional large-scale operating systems including Windows and iOS are too complex and resource-intensive for the majority of IoT applications.

- Event Stream Processing – some IoT applications will generate extremely high data rates that must be analysed in real time. Systems creating tens of thousands of events per second are common, and millions of events per second can occur in some telecom and telemetry situations. To address such requirements, distributed stream computing platforms (DSCPs) have emerged. They typically use parallel architectures to process very high-rate data streams to perform tasks such as real-time analytics and pattern identification.
- IoT Platforms – IoT platforms bundle infrastructure components of an IoT system into a single product. The services provided by such platforms fall into three core categories: (1) low-level device control and operations such as communications, device monitoring and management, security, and firmware updates; (2) IoT data acquisition, transformation and management; and (3) IoT application development, including event-driven logic, application programming, visualization, analytics and adapters to connect to enterprise systems.
- IoT Standards and Ecosystems – ecosystems and standards are not precisely technologies, most eventually materialize as application programming interfaces (APIs). Standards and their associated APIs will be essential because IoT devices will need to interoperate and communicate, and many IoT business models will rely on sharing data between multiple devices and organizations.

Many IoT ecosystems will emerge, and commercial and technical battles between these ecosystems will dominate areas such as the smart home, the Smart City and healthcare. Organizations creating products may have to develop variants to support multiple standards or ecosystems and be prepared to update products during their life span as the standards evolve and new standards and related APIs emerge.

The IERC IoT SRIA addresses these IoT technologies and covers in a logical manner the vision, the technological trends, the applications, the technology enablers, the research agenda, timelines, priorities, and finally summarises in two tables the future technological developments and research needs.

The field of the IoT is based on the paradigm of supporting the IP protocol to all edges of the Internet and on the fact that at the edge of the network many (very) small devices are still unable to support IP protocol stacks. This means that solutions centred on minimum IoT devices are considered as an additional IoT paradigm *without IP to all access edges*, due to their importance for the development of the field.

3.3 IoT Smart Environments and Applications

The IERC vision is that "the major objectives for IoT are the creation of smart environments/spaces and self-aware things (for example: smart transport, products, cities, buildings, rural areas, energy, health, living, etc.) for climate, food, energy, mobility, digital society and health applications" [45].

Today, there is a strong acceleration in the evolution of connected devices, with accelerating scale and scope, as well as higher focus on interoperability. IoT technologies and applications put more and more emphasis on integration of sensors, devices and information systems across industry verticals and organisations to transform operations and enable creation of new business models. IoT technologies focus on gaining new insights from analytics based on data from diverse sources to support decision-making, and improve products, services and experiences for end users. It is envisaged that our environment becomes increasingly "smart" by using this network of connected sensors.

Increasingly complex IoT solutions require more advanced communication platforms and middleware that facilitate seamless integration of devices, networks and applications. In this context, the emergence of IoT platforms with multiple functionalities (i.e. connectivity management, device management, application enablement, etc.) developed for the purpose of supporting and enabling IoT solutions enables rapid development and lower costs by offering standardised components that can be shared across multiple solutions in many industry verticals.

The IoT applications however will gradually move from vertical, single purpose solutions to multi-purpose and collaborative applications interacting across industry verticals, organisations and people, being one of the essential paradigms of the digital economy. Many of those applications still have to be identified and involvement of end-users in this innovation is crucial.

Digital economy enables and conducts the trade of goods and services through electronic commerce on the internet. The digital economy is based on three pillars: supporting infrastructure (hardware, software, telecoms, networks, etc.), e-business (processes that an organisation conducts over computer-mediated networks) and e-commerce (transfer of goods online) [5].

This definition needs to be extended as IoT applications and technologies are more and more embedded in our society. Economic activities classified as "digital economy" are expanding their scale, and are becoming diversified in their transaction forms with many companies in providing product and service hybrids. Intelligent physical goods as part of IoT applications are capable of connecting, capture and producing "smart" data and information for use in digital services without human interventions. In this context, physical equipment has measuring and communication capabilities, data consciousness and processing capabilities and the digital economy will be driven by IoT "system of systems" interactions where new business models and product-service combinations are aligned with customers that are integrating the concept of product-as-a-service and product-as-an-experience.

IoT is expected to boom in many sectors, such as smart buildings and cities, in the energy sector, in safety and security management, transportation, healthcare, farming and many more, thereby bringing huge business opportunities and jobs in those sectors as well as in the enabling industries (data centres, communications and information technology).

The IoT applications are addressing the societal needs and the advancements to enabling technologies such as nanoelectronics and cyber-physical systems continue to be challenged by a variety of technical (i.e., scientific and engineering), institutional, and economical issues.

IERC is focusing on applications chosen as priorities for the next years and the Cluster provides the research challenges for these applications. While the applications themselves might be different, the research challenges are often the same or similar.

Every industry is being disrupted by IoT, the universe of intelligent devices, processes, services, tools and people communicating with each other as part of a global ecosystem. As technology evolves, products, homes, enterprises and entire cities will be continuously connected as presented Figure 3.14. This represents fundamental change for the insurance industry:

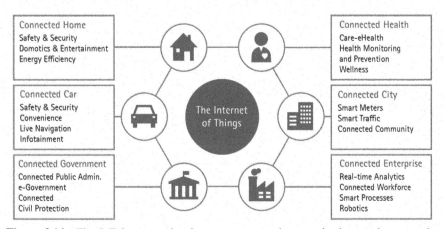

Figure 3.14 The IoT is connecting homes, cars, people, organizations and even entire cities [9].

How are things insured? With what partners? Which services and enabling technologies? The answers to these questions are the first steps toward the development of new and innovative business models. The IoT is driving a connected, as-a-service economy, and traditional insurers must adapt quickly, deciding whether to move up or out. Insurers will need to dramatically reshape their business model, combining insurance with technology, ecosystem services and partners. Insurers are about to become "Insurers of Things" [9].

This new dimension has to be consider for IoT use cases and applications covering various domains and even more when we consider cross-domain applications and implementations.

3.3.1 Wearables

Wearables are integrating key technologies (e.g. nanoelectronics, organic electronics, sensing, actuating, communication, low power computing, visualisation and embedded software) into intelligent systems to bring new functionalities into clothes, fabrics, patches, watches and other body-mounted devices.

These intelligent edge devices are more and more part of integrated IoT solutions and assist humans in monitoring, situational awareness and decision-making. They can provide actuating functions for fully automated closed-loop solutions that are used in healthcare, well-being, safety, security, infotainment applications and connected with smart buildings, energy, lighting, mobility or smart cities IoT applications. Many people already use wearables to monitor their activity level or as a fashion accessory. For example, many of us have a fitbit or a smartwatch.

Creating a seamless user experience is essential for wearable application success. In the future, wearable devices will be more pervasive (e.g. embedded in clothes or pills) and more multifunctional (smartwatches that open doors, start cars and so on) and will become an essential part of people's life.

The IoT applications market in Europe and in the world is moving very fast towards industrial solutions, e.g. smart cities, homes, buildings. The IoT markets have multiple shapes, from simple smart-X devices to complete ecosystems with a full value chain for devices, applications, toolkits and services. Wearables' worldwide market has been identified as the opportunity to materialize what the IoT area has not addressed yet in terms of business creation and commercialization of devices "things", software platforms,

Figure 3.15 Wearables defining priorities for European market.

applications and complete IoT solutions. "Wearables will become the world's best-selling consumer electronics product after smartphones", according to Euromonitor [4]. In the same study the big estimation for sales of wearables are projected to exceed 305 million units in 2020, with a compound annual growth rate (CAGR) of 55 percent during the next five years. Following this big estimation, yet at the Wearables area there is a need for a catalyst that looks for the wider deployment and market uptake of novel/emergent wearables-based IoT applications, technologies and platforms.

The market for wearable computing is expected to grow six-fold, from 46 million units in 2014 to 285 million units in 2018 [36].

Because of wearables are associated to daily life activities and the tendency is to personalise them, following art and design influenced (user-centric) approaches is also crucial. Wearables and its "wear" nature (mobility) will transform diverse sectors such as the healthcare, wellbeing, work safety, public safety and leisure. By involving end users in the creation, the design of

Jewerly   Footwear   bands   Eyewear   Earwear   Wristwear   Vestwear   Straps   Textile

Figure 3.16   Common wearables on the market.

wearables and the identification of services needs, it is expected an exponential growth in the ecosystem for wearables market application.

Wearable technology has been there since early 80's, however the limitation in technology and the high cost on materials and manufacturing generated that wearable ecosystem(s) lost acceptance and stop its grow at that early stage. In todays' technology and economy conditions where technology has evolved and manufacturing cost being reduced Wearable Technology is the best channel for user acceptance and deployments in large user communities. In wearables co-existence with IoT systems and deployed technologies will mark the difference using today's user experience and accelerating tomorrow's user acceptance that is reflected in return on investment by focusing in the most common wearable devices.

Demands in technologies and platforms (supply side) require further work to cope with interoperability, design and arts for user adoption, technology and management and business modelling. In the other hand from users and communities (demand side) it is required to pay attention in reliability of devices, cross-domain operation, and cost reduction and device reusability.

Today's biggest challenges for wearable technology is the reticence to use wearables for privacy or data protection concerns, or the fatigue of using a wearable. In addition, other operational issues also exist such as having the necessary ecosystem in place to support wearable devices, which act as a barrier to deployment, service development or take up. Creating products which meet both end-users need and which create value for the suppliers and users will ensure viable business cases. Wearable devices, which can take or recommend an action based on real time data analysis and perform more than one function (e.g. pain monitoring/treatment that also serves as a security verification that open doors) are more likely to be taken up by different groups of users. They are also more likely to consider them as essential part of their life.

Fitness tracking is the biggest application today and this opens the opportunities for watches that are capable of tracking blood pressure, glucose, temperature, pulse rate and other vital parameters measured every few seconds for a long period of time to be integrated in new kinds of healthcare applications. Glasses for augmented reality can be another future wearable application.

Healthcare industry is taking huge advantage of smart technology for mobile devices and smart wearables is looking to be a big and profitable market. Smart technology that will be the key to the optimal operating of our future society, especially when it comes to healthcare. Some of the smart wearables, already on the market, or in progress engineered for the healthcare industry have the following features [29]:

- Asthma monitoring and management device with companion app currently in design and production phase, offering real time data and alerts when an asthma situation is experienced, offering journaling, treatment plans, displays and tracking and information on the treating of symptoms
- Device attached to a person's back with a companion app, used to lower back pain and treatment with video game like interactions and interface that give the user exercises to do
- Knee brace with companion app giving stability and pain relief using an electrode placed inside of the brace
- Reusable biosensor embedded in a disposable patch with ECG electrodes and accelerometer monitoring heart rate, breathing, temperature, steps, and body position
- Wearable, wireless ECG monitor under development, strapped around the chest to monitor hearth health and health status, with activity tracking monitoring, a companion app, and connection to a cloud based system allowing a doctor to monitoring a patient in real time
- Pill with ingestible sensor technology to be swallowed, powered by the stomach fluids and sending information of your body's physiologic responses and behaviours to a body-worn and disposable patch which can detect heart rate, activity, and rest, and send information to a mobile device. Information if a patient has taken his prescribed medicines at the correct time and how the patient is responding to the therapies
- Smart device helping people to quit smoking by sensing a person's craving for cigarette/nicotine and then deliver medication to curtail the craving, in addition to giving information about quitting and coaching by a companion app
- Smart contact lenses measuring the glucose levels in the wearer's tears, transmitting this information wirelessly to connected smartphones

- Smart contact lenses under development helping restoring the eye's natural autofocus on near objects for people suffering from vision loss occurring with age
- Smart bra and app under development with sensors embedded to sense the conditions and rhythms in breast tissue to alert of the possibility of cancer
- Diabetes sensors placed on the back of the upper arm for 14 days, reading glucose information and transferring this to a companion app, which also give information about the food people should eat, exercise and proper dieting
- Hospital ulcer monitor put on a patient giving the caregiver an alert if the patient is moving around wrong or if they may need some assistance in moving the proper way to prevent ulcer
- Smart watch with medical grade sensors for kids with certain ailments such as epilepsy with real time data sent to a companion app giving alerts and other goals and health information.

The wearable technology market in Europe remains an emerging market that is expanding across numerous sectors and promises to create new markets and deliver important societal benefits. Research from CCS Insight shows that, based on current trajectories, the Global Wearables market is expected to triple by 2019. AIOTI WG07 [65] saw Europe's natural strengths in privacy, data protection as well as in ubiquitous broadband availability enabling Europe to be a strong global competitor in the wearable technologies and solutions sector. If we add Europe's good brand name and talent in style and fashion then we can claim that Europe can be a leader in the market of wearables.

3.3.2 Smart Health, Wellness and Ageing Well

Healthcare and wellness offer unique opportunities for comprehensive IoT implementation. Health care treatments, cost, and availability affect the society and the citizens striving for longer, healthier lives. IoT is an enabler to achieve improved care for patients and providers. It could drive better asset utilization, new revenues, and reduced costs. In addition, it has the potential to change how health care is delivered.

The emergence of Internet of Health (IoH) applications dedicated to citizens health and wellness that spans care, monitoring, diagnostics, medication administration, fitness, etc. will allow the citizens to be more involved with their healthcare. The end-users could access medical records, track the vitals signals with wearable devices, get diagnostic lab tests conducted at home or at the office building, and monitor the health-related habits with Web-based applications on smart mobile devices. The application of IoT in healthcare can improve the access of care to people in remote locations or to those who are incapacitated to make frequent visits to the hospital. It can also enable the prompt diagnosis of medical conditions by measuring and analysing a person's parameters. The medical treatment administered to the person under care can be improved by studying the effect of a therapy and the medication on the patients' vitals.

The IoT healthcare applications require a careful balance between data access and sharing of health information vs. security and privacy concerns. Some information could be shared with a physician, while other type of information, will be not accepted to be provided divulge. For these applications, there is a need to have paradigm shift in human behaviour in order for patients to evolve, adapt and ultimately embrace what the IoT technology can provide, a secure Internet domain that can host all health information and push important health data back to the patient and their healthcare providers [59]. The state of health in a population can be best measured by focusing on metabolic syndromes with a set of clear and staged health actions attached to it in order to fight the consequences of such modern lifestyle. If not changed, this lifestyle often results in an early progression of those diseases (as shown in Figure 3.17) [63].

The population of people over 60 is growing at a faster rate than the rest of the population. Unlike previous generations, more seniors will stay at home. In the future IoT technology might allow older people to retain independence with a choice to keep family informed when help is needed. Silver Economy is defined as "an environment in which the over-60 interact and thrive in the workplace, engage in innovative enterprise, help drive the marketplace as consumers and lead healthy, active and productive lives" [71]. There are three groups in the ageing population, depending on their health, i.e. active, fragile and dependent while each of these groups have their own need patterns. At country level differences in needs patterns exist, i.e. depending on the local environment, with the existence of models for care, governmental policy and needs at European geographical levels, i.e. Nordic, Anglo-Saxon, Continental, South-European and Eastern-European. The Silver Economy is related to concepts such as "active and health ageing", "ambient assisted living", "e-health", "age management", "smart care" etc.

Chronic Quadrangle: Behaviour intensive diseases with deferred consequences

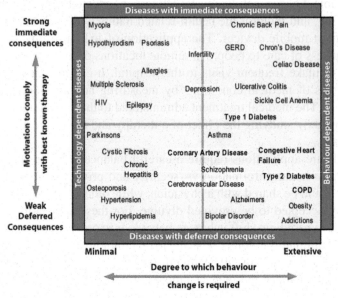

Figure 3.17   Chronic quadrangle.

and depends on the perspective taken or challenge/solution emphasised, using different taxonomies.

Demographic change, the rising incidence of chronic disease, unmet demand for more personalised care, and cost pressure are trends requiring a new, integrated approach to health and social care. Such integration – if brought about in the right manner – has the potential to improve both the quality, security and the efficiency of care service delivery. Potentially this can be to the benefit of all: beginning with elder people in need of care and their family and friends, and including care professionals, service provider organisations, payers and other governance bodies. Within this ongoing change process, the challenge is how to adopt relevant and secure IoT technologies to realise care integration and avoid that telecare, telehealth and other IoT applications in this field remain locked up in segregated silos, mirroring the overall situation of today. In order to capture all the complexity of the ambient assisted living (AAL) market scenario, the previous definition was taken into account as a starting point but have also taken into account a technology view, based on the technology stack supporting the AAL solutions.

IoT applications are pushing the development of platforms for implementing AAL systems that will offer services in the areas of assistance to carry out daily activities, health and activity monitoring, enhancing safety and security, getting access to medical and emergency systems, and facilitating rapid health support.

The main objective is to enhance life quality for people who need permanent support or monitoring, to decrease barriers for monitoring important health parameters, to avoid unnecessary healthcare costs and efforts, and to provide the right medical support at the right time.

The IoT plays an important role in healthcare applications, from managing chronic diseases at one end of the spectrum to preventing disease at the other.

The smart living environments at home, at work, in public spaces should be based upon integrated systems of a range of IoT-based technologies and services with user-friendly configuration and management of connected technologies for indoors and outdoors.

These systems can provide seamless services and handle flexible connectivity while users are switching contexts and moving in their living environments and be integrated with other application domains such as energy, transport, or smart cities. The advanced IoT technologies, using and extending available open service platforms, standardised ontologies and open standardised APIs can offer many of such smart environment developments.

The IoT technology not only overcomes the inconvenience of distance, but also provides people with greater choice and control over the time and the place for monitoring their condition, increasing convenience and making their conditions more manageable. At the same time, it also reduces some of the pressures on clinics and acute hospitals. IoT could make a significant contribution to the management of a number of chronic conditions, heart failure, hypertension, asthma, diabetes and can be integrated with other living environments domains such as mobility, home/buildings, energy, lighting, cities.

Many elderly people are adopting technology more than ever, and in the process, they face unique barriers to usage because they previously had not used them in work situations and commonly have physical limitations that make using computer and the Internet more difficult. The improvement in the IoT technology and user interfaces can lower the barriers and help the elderly people to adopt the technology since many of these people are enthusiastic and express strong openness to learning.

As the population ages, and as the digital health field expands, IoT technologies addressing the unique challenges of aging in place is becoming a reality.

Many elderly people want to age in place and need to be as independent as possible, while the IoT technology provides cognitive aids for independent living. Old people with Alzheimer's, dementia, or memory loss receive help with tasks through cueing, scheduling assistance and finance safety for seniors by on and off switches for caregivers or relatives to help aging people manage their money by blocking purchases, setting spending limits, sending alerts about suspect charges, etc. IoT activity sensors monitor movements in the home and medicine boxes give medication reminders, keep track of steps, and include an emergency button.

The IoT allows building up an archive of patient behaviour in their own home that will enable local analytics to produce probability curves to predict usual and unusual behaviour. Using this, a more accurate prediction of unusual behaviour can be detected that is used to trigger alerts to patients, family and carers, while helping elderly patients stay out of hospital (and thus significantly reduce the cost of hospital admissions).

In this context, there is a need for fundamental shift in the way we think about older people, from dependency and deficit towards independence and well-being. Older people value having choice and control over how they live their lives and interdependence is a central component of older people's well-being. They require comfortable, secure homes, safe neighbourhoods, friendships and opportunities for learning and leisure, the ability to get out and about, an adequate income, good, relevant information and the ability to keep active and healthy. They want to be involved in making decisions about the questions that affect their lives and the communities in which they live. They also want services to be delivered not as isolated elements, but as joined-up provision, which recognises the collective impact of public services on their lives. Public services have a critical role to play in responding to the agenda for older people.

Within this ongoing change process, advanced IoT technologies provide a major opportunity to realise care integration. At the same time, telecare, telehealth and other IoT applications in this field also remain locked up in segregated silos, mirroring the overall situation.

These IoT technologies can propose user-centric multi-disciplinary solutions that take into account the specific requirements for accessibility, usability, cost efficiency, personalisation and adaptation arising from the application requirements.

3.3.3 Smart Clothing

Smart textile, e-fabrics, smart clothing will be produced in all kinds of types and with different features and outlooks and in many cases will embed the features and functionalities of wearable devices of today. The common factor is that smart textiles are made to observe to the wearer, and to react to environmental conditions including chemical, mechanical, electrical, chemical, and magnetic, etc. Intelligent fabrics have digital components, sensors, actuators, circuits, and computers embedded in them to collect process and output data in different ways.

Smart clothing will include many features and different smart solutions are expected on the market in the next years [30, 31]:

- Smart shirt with app, keeping information in 3D showing if too much pressure is put on a certain part of the body, keeping track of your performance, giving information to prevent getting injured while training, with real time feedback
- Health related smart shirt measuring heart rate, breathing rate, sleep monitoring, workout intensity measurements
- Bio sensing silver fibres woven into the shirt
- Clothing to track the amount of calories burned
- Clothing to track movement intensity during workout
- Compression fabric that aids in blood circulation and with muscle recovery
- Body monitor sensors – embedded micro sensors throughout the shirt keeping track of temperature, heart beat and heart rate, and the speed and intensity of your workouts
- Shirt able to keep the measured biometrics information by using a small black box woven into the shirt
- Clothing with moisture control and odour control
- Smart shirts can be used in hospitals for monitoring heart beat and breathing in patients
- Baby monitoring – baby garment telling if the baby is sleeping and monitoring the baby's vital signs
- Baby outfit with sensors and a small monitor on it
- Smart socks for baby, monitoring the baby's breath with alert features
- Eco-friendly solar garments as it harnesses the energy of the sun and enables the wearer to charge the owner's phone, music players, and other powered electronic devices

- Adaptive survival clothing that uses moisture and temperature regulation properties of wool to adapt the human body to normal, non-threatening conditions.

The combination of these "devices" embedded in the clothing with other IoT devices that are monitoring the environment will create new opportunities, new use cases, and business models across various sectors.

3.3.4 Smart Buildings and Architecture

Buildings consume 33% of world energy, this figure grows to 53% of world electricity, and it will continue to grow in the future. As a result, buildings have an important weight in regards to the energy challenge.

Improving life of the occupants implies many aspects including comfort with light, temperature, air quality, having access to services facilitating life inside the building, adapting the behaviour to the needs of the occupants. There is also a direct economic interest to do it as it is recognized that productivity level is connected to the comfort level.

For being energy efficient, the consumption can be optimized locally while taking into account the needs of the occupants and the hosted processes. Buildings can also produce energy from different sources such as Photovoltaic panels and store energy for future usage. This energy can be used internally or given back to the grid. In addition, buildings are not isolated islands but part of larger ecosystem at the district or even city level. The energy price can change over time and have an impact on the energy optimization. It could happen also that the optimization is better driven at a more global level, set of buildings or district for instance. In the smart building implementations, it is necessary to simplify the management, control and maintenance of buildings during the whole life cycle, starting from the design phase. This should lead to much better process efficiency while driving down the operation costs (OPEX).

As a result there is a strong need to leverage on technology and IoT for making buildings smarter, improve life of the occupants (personal or at work), make the buildings more energy efficient, and facilitate the management and maintenance of the building during its whole life cycle. This has to be done not only with the new constructed buildings but also with the existing ones through adequate retrofit solutions. It is important to keep in mind that new construction represents only 2% of the total installed base each year.

The different ingredients of IoT, connectivity, control, cloud computing, data analytics, can all contribute to make smarter buildings (offices, industrial, residential, tertiary, hotels, hospitals, etc.):

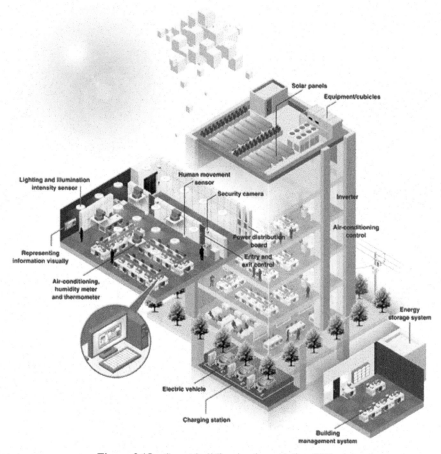

Figure 3.18 Smart building implementation [72].

- Connected to the grid ("smart grid ready")
- Connected to the Smart City
- Energy efficient while taking care of the comfort of the occupants
- Adaptable to the changing needs of the occupants over time
- Providing services for a better life of the occupants
- Easy to maintain during the whole life cycle at minimal cost

The solutions focus primarily on environmental monitoring, energy management, assisted living, comfort, and convenience. The solutions are based on open platforms that employ a network of intelligent sensors to provide information about the state of the home. These sensors monitor systems such

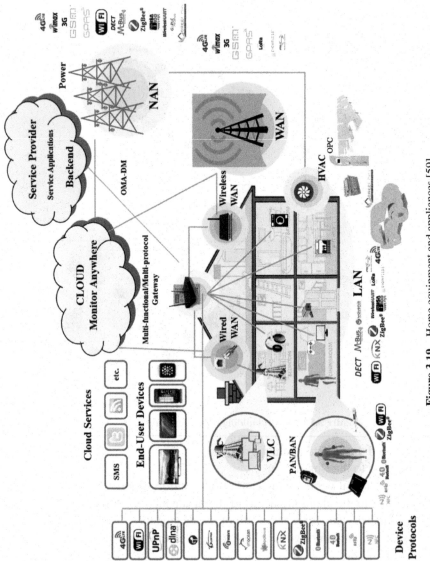

Figure 3.19 Home equipment and appliances [59].

as energy generation and metering; heating, ventilation, and air-conditioning (HVAC); lighting; security; and environmental key performance indicators.

The networking aspects are bringing online streaming services or network playback, while becoming a mean to control of the device functionality over the network.

Integration of cyber-physical systems (CPS) both within the building and with external entities, such as the electrical grid, requires stakeholder cooperation to achieve true interoperability. Maintaining security will be a critical challenge to overcome in smart buildings IoT applications [71].

In the IoT ecosystems, the collaboration among various stakeholders to optimise the smart buildings allow operators of buildings to find ways to conserve energy for both environmental and economic reasons, while architects and builders, are trying to make new buildings as "green" as possible.

IoT technologies are extending today's building automation and transforming the smart buildings and facilities through IoT platforms providing intelligence, security, modularity, and intuitive interfaces that allow autonomous operations. The evolution of building system architectures includes an adaptation level that will dynamically feed the automation level with control logic, i.e. rules, using algorithms and rules as Web resources in a similar way as for sensors and actuators.

The market sizing and opportunities for smart commercial buildings; is increasing and Memoori report "The Internet of Things in Smart Buildings 2014 to 2020" [33] makes an objective assessment of the market for IoT Technologies, Networks and Services in Buildings 2014 to 2020. Market figures indicate that the overall market for systems in buildings will grow from \$110.9Bn in 2014 to \$181.1Bn in 2020, with Physical Security, Lighting Control and Fire Detection and Safety representing the three largest segments. In order to calculate the technical market potential for the IoT in Buildings.

The report has assessed the additional cost requirement of adding connectivity through sensors to existing or newly installed building systems, as well as projecting the growth in related network hardware and IoT data services that the IoT in Buildings would enable to generate. The report therefore projects that the global market for the IoT in Buildings will rise from \$22.93Bn in 2014 to over \$85Bn in 2020. In this context, the following estimates are made:

- Overall connectivity penetration rates across all building systems are at only around 16%. This connectivity penetration rate will rise steadily over the coming years, but mainstream penetration, i.e. 50% of all building systems devices connected, is unlikely to be achieved before 2025.

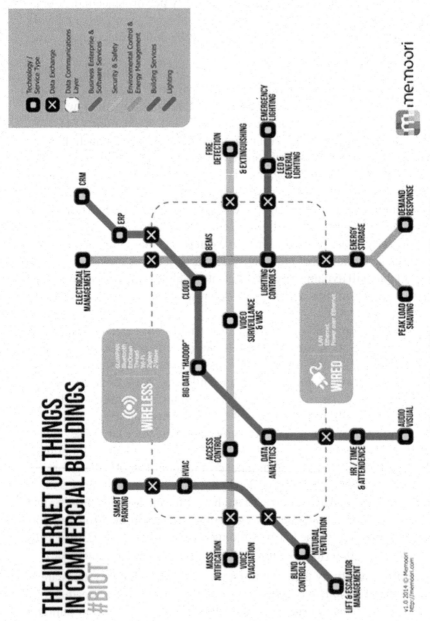

Figure 3.20   IoT in commercial buildings [33].

- The networking and related services segment of the market will show a steady growth of 22.6% CAGR rising from \$9.53Bn in 2014 to \$32.43Bn in 2020 which represents 37% of overall revenues by 2020. Similar to the market for connectivity hardware, effective network deployment to keep up with the rising bandwidth demands of the IoT in Buildings will be crucial to the effective delivery of services and the management of data flows.

The concept of "Internet of Building" that integrates the information from multiple intelligent building management systems and optimise the behaviour of individual buildings as part of a larger information system. These systems are used by facilities managers in buildings to manage energy use and energy procurement and to maintain buildings. It is based on the infrastructure of the existing Intranets and the Internet, and therefore utilises the same standards as other IT devices. Reductions in the cost and increased reliability of IoT applications using wireless technologies for monitoring and control are transforming building automation, by making the maintenance of energy efficient healthy productive workspaces in buildings increasingly cost effective [50].

IoT technologies and applications used across the buildings and architecture sector need to be integrated with applications in other sectors. The value in "Internet of Buildings" is as much in the edge devices and the data collected, exchanged and processed. Collecting, exchanging and processing data from building services and equipment provides a granular view of how each building is performing, allowing the development of building systems that collect, store and analyse data at the edge and in the cloud, providing better operational efficiency and integration with IoT platforms and applications across various sectors. These efforts will cover the following domains of research.

- IoT architecture and IoT platforms to address smart buildings and architecture monitoring and control strategies and integrate monitors/controls from edge sensors/actuators devices to the data exchange and processing.
- Communication technologies and infrastructures required for IoT buildings applications and their integration with applications and IoT platforms across various consumer and industrial sectors.
- Hardware/software, machine learning and analytics approaches supporting real-time interoperable distributed decision support monitoring and control in heterogeneous environments.
- New developments in the smart buildings addressing business models, applications, IoT technology, interoperability at various levels and frameworks, regulation and law, etc.

3.3.5 Smart Energy

Future energy supply will be largely based on various renewable resources and this source of energy will influence the energy consumption behaviour, demanding an intelligent and flexible electrical grid which is able to react to power fluctuations by controlling electrical energy sources (generation, storage) and sinks (load, storage) and by suitable reconfiguration. The functions are based on networked intelligent devices (appliances, micro-generation equipment, infrastructure, consumer products) and grid infrastructure elements, largely based on IoT concepts.

The energy grid development requires a number of features as listed below in order to implement the vison of the smart grid concept.

- It will integrate traditional and emerging power sources and make the delivery of energy cleaner, safer, and more economical.
- Operators will have the transparency and visibility to monitor and analyse the flow of energy, and two-way communication with consumers' smart meters to analyse consumption patterns.
- Intelligent devices that collect and analyse massive volumes of data will enable operators to plan for contingencies for variable resources.
- Smart IoT devices will manage the distribution of energy based on real-time data and situational awareness, as opposed to historical data patterns.
- Predictive maintenance capabilities will alert operators when a component needs attention or repair, reducing the need for ongoing inspections.
- Adaptive analytics will enable systems to automatically balance energy loads to reduce stress and prevent overheating.

The high number of distributed small and medium sized energy sources and power plants can be combined virtually ad hoc to virtual power plants. Using this concept, areas of the grid can be isolated from the central grid and supplied from within by internal energy sources such as photovoltaics on the roofs, block heat and power plants or energy storages of a residential area.

IoT is expected to facilitate the deployment of new smart energy apps within energy stakeholders ICT (generation and retail companies, Grid and market operators, new load aggregators) bringing new options for real-time control strategies across energy asset portfolios for faster reactions to power fluctuations. These new technologies should combine both centralised and decentralised approaches integrating all energy generation (generation, storage) and load (demand responsive loads in residential, buildings and industries as well as storage and electrical vehicles) through interconnected real-time energy markets. IoT should also improve the management of asset

Figure 3.21 Smart grid concept [49, 79].

performance through more accurate estimations of asset health conditions and deployment of fact based preventive maintenance.

These new smart energy apps will largely be based on the networking of IoT intelligent devices embedded within Distributed Energy Resources (DER) spread across the energy system such as consumer appliances, heating and air conditioning, lighting, distributed generation and associated inverters, grid edge and feeder automation, storage and EV charging infrastructures. While energy systems have historically been controlled through single central dispatch strategies with limited information on grid edge and consumers behaviours, energy systems are now characterized by rapidly growing portfolios of DER structured through several layers of control hierarchies interconnecting the main grid down to microgrids within industries and communities, nanogrids at building level and picogrids at residential scale.

Moreover as most of DER have diffused within end-user premises, new transactive control approaches are required to facilitate their coordination at various scales of the Grid system through real-time pricing strategies. Furthermore aggregators and energy supply companies have started to develop new flexibility offers to facilitate DER coordination virtually through ad hoc virtual power plants raising new connectivity, security and data ownership challenges.

Meanwhile climate change has also recently exposed grids to new extreme weather conditions requiring reconsidering Grid physical and ICT architectures to allow self-healing during significant disasters while taking advantage of distributed generation and storage to island critical grid areas (hospital, large public campus) and maintain safe city areas during emergency weather conditions.

Integration of cyber-physical systems engineering and technology to the existing electric grid and other utility systems is a challenge. The increased system complexity poses technical challenges that must be considered as the system is operated in ways that were not intended when the infrastructure was originally built. As technologies and systems are incorporated, security remains a paramount concern to lower system vulnerability and protect stakeholder data [71]. These challenges will need to be address as well by the IoT applications that integrate heterogeneous cyber-physical systems.

A new report by Mercom Capital Group indicates that smart grid, battery and storage, as well as energy efficiency companies raised up to US$1.7bn in VC funding in 2015. The report which examines mergers and acquisition activity in the smart grid, battery/storage, and energy efficiency sectors, revealed that the smart grid sector raised US$425 million across 57 deals in 2015, in comparison to US$384 million over 74 deals in the previous year (2014) [79].

The energy grid is expected to be the implementation of a kind of "Internet" in which the energy packet is managed similarly to the data packet – across routers and gateways, which autonomously can decide the best pathway for the packet to reach its destination with the best integrity levels. In this respect, the "Internet of Energy" concept is defined as a network infrastructure based on standard and interoperable communication transceivers, gateways and protocols that will allow a real time balance between the local and the global generation and storage capability with the energy demand.

The Internet of Energy (IoE) concept is defined as a network infrastructure based on standard and interoperable communication nodes that will allow the end-to-end real time balance between the local and the central generation, responsive demand and storage. It will allow units of energy to be transferred when and where it is needed. Power consumption monitoring will be performed on all levels, from local individual devices up to national and international level [78].

Considering the fast diffusion of energy resources on end user premises – becoming prosumers-, the new IoT platform considered will also allow a high level of consumer awareness and involvement through community benchmarking.

Electro mobility requiring the rapid deployment of charging infrastructures adding significant constraints to power grids; EVs will be considered as integral element of future smart energy systems acting as a power load as well as moveable energy storage linked through IoT technologies. EVs will require to transact with the Energy system according to their charge status, usage schedule and energy price which itself will depend on abundance of renewable energy available at a certain time in the energy system. This should ultimately allow monitoring the carbon footprint of all mobility services from wells to wheels.

Latencies are critical when talking about electrical control loops. Even though not being a critical feature, low energy dissipation should be mandatory. In order to facilitate interaction between different vendors' products the technology should be based on a standardized communication protocol stack.

When dealing with a critical part of the public infrastructure, data security is of the highest importance. In order to satisfy the extremely high requirements on reliability of energy grids, the components as well as their interaction must feature the highest reliability performance.

IoT applications in the energy sector go beyond one industrial sector. Energy, mobility and home/buildings sectors will have to share data through energy gateways that will control the transfer of energy and information.

Flexible data filtering, data mining and machine learning procedures as well as new generation IoT platforms are necessary to handle the high amount of raw data provided by billions of data sources while guaranteeing resiliency, security as well as end user data protection. System and data models need to support the design of real-time decision support systems, which guarantee a reliable and secure operation of vital energy infrastructures.

The future research challenges will cover the following areas:

- ICT/IoT architectures and IoT platforms to revisit grid control strategies and integrate hierarchical controls from energy nodes with sensors through ranges of aggregation structures (pico, nano and micro energy systems).
- Novel communication infrastructures required at each level of these grid nodes to meet necessary Service level agreements for each of the energy service considered (energy efficiency, grid ancillary services, grid resiliency, etc. . . .).
- New software/smart data and machine learning approaches supporting real-time distributed decision support/transactive controls in highly volatile environments.

- New apps for energy prosumer feedback facilitating smooth real-time energy transactive controls in daily lives leveraging consumer ICT (mobile, TVs, vehicle, IoT, etc. . .).
- IoT end-to-end security framework approach and privacy, trust and safety in order to secure the grid from hackers and acts of cyber-sabotage. Security needs to be built into every device starting at the base of the software stack.
- Providing intelligent solution for connecting and protecting legacy systems (the older, aging parts of the existing energy infrastructure) by building secure Internet gateways that enable cloud-based central control systems to collect local intelligence data from the systems while blocking attacks.
- Embedding intelligence into the energy systems with smart energy devices that deliver manageability, security, and connectivity, while driving down the cost of development and deployment.
- Privacy by design of the energy systems that will assure that the data generated by using the monitoring systems will not expose sensitive customer information. This requires that the same security principals that apply to the energy enterprise will also be applied at the consumer level.

3.3.6 Smart Mobility and Transport

The connection of vehicles to the Internet gives rise to a wealth of new possibilities and applications which bring new functionalities to the individuals and/or the making of transport easier and safer. In this context the concept of Internet of Vehicles (IoV) [78] connected with the concept of Internet of Energy (IoE) represent future trends for smart transportation and mobility applications.

At the same time creating new mobile ecosystems based on trust, security and convenience to mobile/contactless services and transportation applications will ensure security, mobility and convenience to consumer-centric transactions and services.

Representing human behaviour in the design, development, and operation of cyber-physical systems in autonomous vehicles is a challenge. Incorporating human-in-the-loop considerations is critical to safety, dependability, and predictability. There is currently limited understanding of how driver behaviour will be affected by adaptive traffic control cyber-physical systems. In addition, it is difficult to account for the stochastic effects of the human driver in a mixed traffic environment (i.e., human and autonomous vehicle drivers) such as that found in traffic control cyber-physical systems. Increasing integration calls for security measures that are not physical, but more logical while still ensuring there will be no security compromise. As cyber-physical systems become more complex and interactions between components increases, safety and security will continue to be of paramount importance [71]. All these elements are of the paramount importance for the IoT ecosystems developed based on these enabling technologies.

Self-driving vehicles today are in the prototype phase and the idea is becoming just another technology on the computing industry's parts list. By using automotive vision chips that can be used to help vehicles understand the environment around them by detecting pedestrians, traffic lights, collisions, drowsy drivers, and road lane markings. Those tasks initially are more the sort of thing that would help a driver in unusual circumstances rather than take over full time. But they're a significant step in the gradual shift toward the computer-controlled vehicles that Google, Volvo, and other companies are working on [56].

These scenarios are, not independent from each other and show their full potential when combined and used for different applications.

Technical elements of such systems are smart phones and smart vehicle on-board units, which acquire information from the user (e.g. position, destination and schedule) and from on-board systems (e.g. vehicle status, position, energy usage profile, driving profile). They interact with external systems (e.g. traffic control systems, parking management, vehicle sharing managements, electric vehicle charging infrastructure).

The concept of Internet of Vehicles (IoV) is the next step for future smart transportation and mobility applications and requires creating new mobile ecosystems based on trust, security and convenience to mobile/contactless services and transportation applications in order to ensure security, mobility and convenience to consumer-centric transactions and services.

Smart sensors in the road and traffic control infrastructures need to collect information about road and traffic status, weather conditions, etc. This requires robust sensors (and actuators) which are able to reliably deliver information to the systems mentioned above. Such reliable communication needs to be based on IoT communication, which consider the timing, safety, and security constraints. The integration of the communication gateway into vehicles is presented in Figure 3.23. The expected high amount of data will require sophisticated data mining strategies. Overall optimisation of traffic flow and energy usage may be achieved by collective organisation among the individual vehicles.

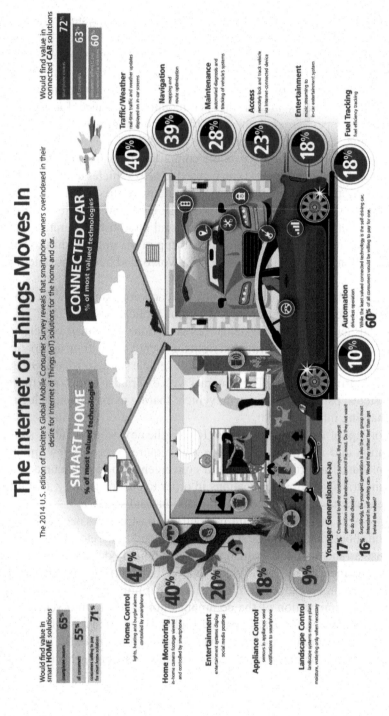

Figure 3.22 Home and vehicle IoT solutions [55].

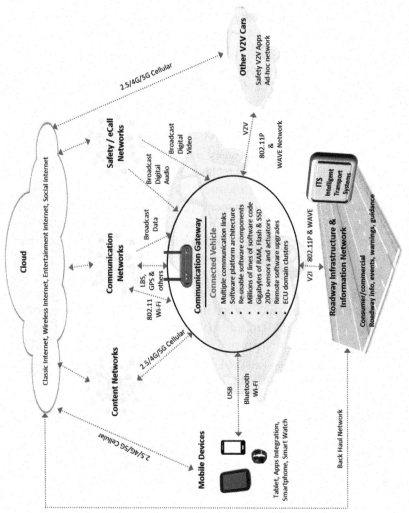

Figure 3.23   Vehicle integrated IoT communication platform.

When dealing with information related to individuals' positions, destinations, schedules, and user habits, privacy concerns gain highest priority. They even might become road blockers for such technologies. Consequently, not only secure communication paths but also procedures which guarantee anonymity and de-personalization of sensible data are of interest.

Connectivity will revolutionize the environment and economics of vehicles in the future: first through connection among vehicles and intelligent infrastructures, second through the emergence of an ecosystem of services around smarter and more autonomous vehicles.

In this context the successful deployment of safe and autonomous vehicles (SAE[1] international level 5, full automation) in different use case scenarios, using local and distributed information and intelligence is an important achievement. This is based on real-time reliable platforms managing mixed mission and safety critical vehicle services, advanced sensors/actuators, navigation and cognitive decision-making technology, interconnectivity between vehicles (V2V) and vehicle to infrastructure (V2I) communication. There is a need to demonstrate in real life environments (i.e. highways, congested urban environment, and/or dedicated lanes), mixing autonomous connected vehicles and legacy vehicles the functionalities in order to evaluate and demonstrate dependability, robustness and resilience of the technology over longer period of time and under a large variety of conditions.

The introduction of the autonomous vehicles enables the development of service ecosystems around vehicles and multi-modal mobility, considering that the vehicle includes multiple embedded information sources around which information services may be constructed. The information may be used for other services (i.e. maintenance, personalised insurance, vehicle behaviour monitoring and diagnostic, security and autonomous cruise, etc.).

The emergence of these services will be supported by open service platforms that communicate and exchange information with the vehicle embedded information sources and to vehicle surrounding information, with the goal of providing personalised services to drivers. Possible barriers to the deployment of autonomous vehicles and ecosystems are the robustness sensing/actuating the environment, overall user acceptance, the economic, ethical, legal and regulatory issues.

The integration of the interconnected and intelligent intra vehicle communication systems and the vehicle to infrastructure into the overall IoT service platforms will offer the possibility to develop new applications and

---

[1] Society of Automotive Engineers, J3016 standard.

services it is expected that 80% of vehicles in Europe will be two-way connected by 2018. This offer the possibility to combine the vehicle to infrastructure communication and integration with service providers with intermodal vehicle navigation applications and navigation routes based on real-time information. IoT applications for vehicle sharing and the use of transport city fleets (EVs for transport of goods and persons) are part of the deployment of new IoT technologies and related IoT ecosystems. These will open the stepwise rollout of autonomous driving technologies and the linkages of these technologies with shared-use business models and issues relating to the regulatory framework and consumer trust.

For autonomous vehicle applications, computing at the edge of the mobile network will be used for processing the data locally and provide services in real time.

Data transmission costs and the latency limitations of mobile connectivity pose challenges to autonomous vehicle IoT applications that cannot rely only on cloud computing.

Mobile edge computing enables IoT applications to deliver real-time and context-based mobile moments to users of IoT solutions.

In IoT applications involving autonomous vehicles a combination of cloud and mobile edge computing technologies have to be consider by analysing the following:

- Cloud, mobile edge and IoT are increasingly intertwined and used together to improve IoT application experiences. IoT solutions gain functionality through cloud services, which in turn open access to third-party companies and up-to-date information.
- Mobile connectivity for real time autonomous systems create challenges for cloud-enabled IoT solutions since latency limitations affects user experiences in the IoT real time applications context.
- Mobile edge computing assure the real time network connectivity, location and context information. The technology gives access to "near edge" computing capabilities and a cloud like service environment close to the users and edge devices.
- Mobile edge computing is a component of the network infrastructure for blockchain, since the replication of "blocks" via devices can be implemented at the edge.

3.3.7 Industrial IoT and Smart Manufacturing

The role of the IoT is becoming more prominent in enabling access to devices and machines, which in manufacturing systems, were hidden in well-designed silos. This evolution will allow the IT to penetrate further the digitized manufacturing systems. The IoT will connect the factory to a completely new range of applications, which run around the production. This could range from connecting the factory to the smart grid, sharing the production facility as a service or allowing more agility and flexibility within the production systems themselves. In this sense, the production system could be considered one of the many Internets of Things (IoT), where a new ecosystem for smarter and more efficient production could be defined.

The evolutionary steps towards smart factory require enabling access to external stakeholders in order to interact with an IoT-enabled manufacturing system that is formed of connected industrial systems that communicate and coordinate their data analytics and actions to improve performance and efficiency and reduce or eliminate downtime. These stakeholders could include the suppliers of the productions tools (e.g. machines, robots), as well as the production logistics (e.g. material flow, supply chain management), and maintenance and re-tooling actors. The manufacturing services and applications do not need to be defined in an intertwined and strictly linked manner to the physical system, but rather run as services in a shared physical world. Adopting the industrial IoT requires a change in the way stakeholders design and augment their industrial systems in order that the IoT industrial systems are adaptive and scalable through software or added functionality that integrates with the overall solution.

Industrial IoT applications are using of the data available, business analytics, cloud services, enterprise mobility and many others to improve the industrial processes. These technologies include big data and business analytics software, cloud services, embedded technology, sensor networks/sensing technology, wireless communication, mobility, security and ID recognition technology, wireless network and standardisation. Security is very important in industrial IoT applications that are processing the information from tens of thousands of edge devices nodes. Faulty data injected into the system has the potential to be as damaging as data extracted from the systems via data breach.

The convergence of microelectronics and micromechanical parts within a sensing device, the ubiquity of communications, the rise of micro-robotics, the customization made possible by software will significantly change the world of manufacturing. In addition, broader pervasiveness of telecommunications in many environments is one of the reasons why these environments take the shape of ecosystems.

The future IoT developments integrated into the digital economy will address highly distributed IoT applications involving a high degree of distribution, and processing at the edge of the network by using platforms that that provide compute, storage, and networking services between edge devices and computing data centres.

IoT applications integrate sensors/actuators and cyber-physical systems offering new opportunities for new combinations of virtual, digital, physical and mechanical work. The IoT and Industrial IoT are currently underlying the far-reaching integration of Information Technology (IT: conventional computers, operating systems, networking components and software platforms.) and Operational Technology (OT: industrial control system and networks, hardware and software that detects or causes a change through the direct monitoring and/or control of physical devices, processes and events in the enterprise) [12, 13].

Some of the main challenges associated with the implementation of cyber-physical systems in include affordability, network integration, and the interoperability of engineering systems.

Most companies have a difficult time justifying risky, expensive, and uncertain investments for smart manufacturing across the company and factory level. Changes to the structure, organization, and culture of manufacturing occur slowly, which hinders technology integration. Pre-digital age control systems are infrequently replaced because they are still serviceable. Retrofitting these existing plants with cyber-physical systems is difficult

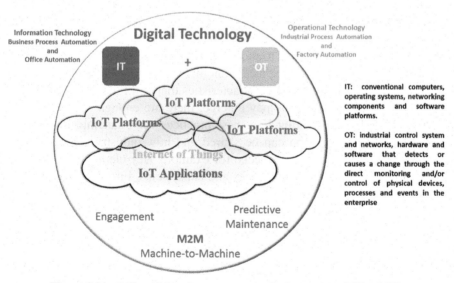

Figure 3.24   IoT providing the core structure for integration of IT and OT.

and expensive. The lack of a standard industry approach to production management results in customized software or use of a manual approach. There is also a need for a unifying theory of non-homogeneous control and communication systems [71].

The industrial IoT is implemented in various forms, one is called Internet of Things, Services and People (IoTSP) [27] were the focus is to develop and enhance process control systems, communications solutions, sensors and software for the IoTSP. These technologies enable the customers in industries, utilities and infrastructure to analyse their data more intelligently, optimize their operations, boost their productivity, and their flexibility. IoTSP is advancing by helping the IoT stakeholders and customers to develop their existing technologies, while keeping sight of our enduring commitment to safety, reliability, cyber security and data privacy. Developing and improving process control system, communication solutions, sensors and software used in IoTSP provide new value for the customers. With these technologies, the customers in industry, utility, transportation and infrastructure can benefit from smart data analysis, optimized operation, and higher productivity and flexibility.

3.3.8 Smart Cities

Cities all over the world, from small regional communities to global mega hubs and from cities with an ancient core to brand new developments, are currently working on 'Smart City' initiatives to make them more efficient, sustainable, and more attractive to citizens and businesses and to encourage economic growth. There are many obstacles to successful implementation of these plans, and translating solutions from one place to another is difficult. While every city on earth is unique and has its own characteristics that will impact why, how and which Smart City solutions may emerge, there are enough similarities for it to be worth investigating how best practices for financing, design, implementation and operation can be shared and how industry can re-use experience gained from earlier projects, for example. Key elements include interoperability of data between devices and subsystems, information flows between project partners, financing, risk management, etc. [57].

A Smart City is defined as a city that monitors and integrates conditions of all of its critical infrastructures, including roads, bridges, tunnels, rail/subways, airports, seaports, communications, water, power, even major buildings, can better optimize its resources, plan its preventive maintenance activities, and monitor security aspects while maximizing services to its citizens. Emergency response management to both natural as well as man-made challenges to the system can be focused and rapid. With advanced monitoring systems and built-in smart sensors, data can be collected and evaluated in real time, enhancing city management's decision-making [69].

There are a number of key elements needed to form a Smart City, and some of these are smart society, smart buildings, smart energy, smart lighting, smart mobility, smart water management etc. ICT forms the basic infrastructure; varying from sensors, actuators and electronic systems to software, Data, Internet and Cloud, Edge/fog and Mobile Edge computing. ICT is applied to improve these systems of systems building up a Smart City, making them autonomous and interoperable, secure and trusted. The interaction of the systems and the connectivity strongly depend on the communication gateway connecting the edge element data from sensors, actuators, and electronic systems to the Internet, managing- and control systems and decision programs.

An illustrative example of a Smart City model is presented in Figure 3.25 [57]. This model has a mostly technical view, concentrating on how (sub) systems interact with each other supported by telecommunications and information technology. The city is divided into the built environment (including homes, offices and shops and the devices within them), infrastructure-based sectors (e.g. energy and waste) and service-based sectors (e.g. healthcare and education). There is possible interaction between elements within any of these subsystems as well as between subsystems. Smart city infrastructure sectors, such as telecommunications, information technology and electronics, enable and support this interaction. A common theme in the example Smart City models is the use of sensors to collect data from the city, which, through platforms, can be combined, stored, analysed and displayed. This provides decision support for actors in the city who can then act and make changes, the effect of which can in turn be measured [57]. The Smart City is not only the integration and interconnection of intelligent applications, but also a people-centric and sustainable innovation model that is using communication and information technology and takes advantage of the open innovation ecology of the city and the new technologies such as IoT, cloud computing, data analytics, human-human, human-machine, machine-infrastructure, machine-environment interaction.

A Smart City is a developed urban area that creates sustainable economic development and high quality of life by excelling in multiple key areas: economy, mobility, environment, people, living, and government [77].

Identifying or developing sets of Key Performance (KPI) and other indicators to gauge the success of Smart City ICT deployments. KPIs are required

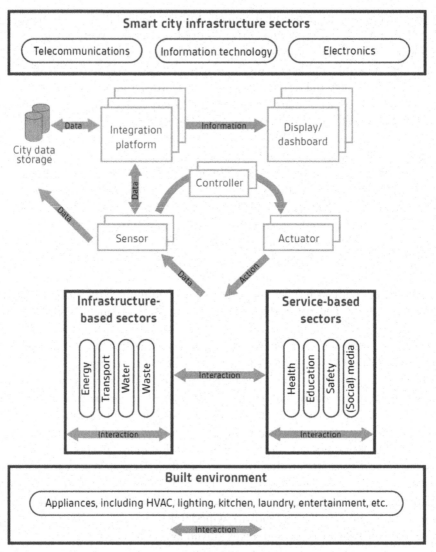

Figure 3.25   Smart City model – technical view [57].

to provide performance as seen from different viewpoints, such as those: of residents/citizens (reliability, availability, quality and safety of services, etc.); of community and city managers (operational efficiency, resilience, scalability, security, etc.); and of the environment (climate change, biodiversity, resource efficiency, pollution, recycling rates/returns). The indicators appropriate for one city or context may not be the same for others. As such, there should also be standardized guidance for city managers on selecting and using KPIs appropriate to their particular situation. Requirements for standardized risk assessment methodologies for critical infrastructure dependencies across organisations and sectors [58].

3.3.8.1 Open Data and Ecosystem for Smart Cities

As main areas of application, smarter cities plays a relevant role, not only because the impact in re-using and re-purposing technology that is necessary (the number of deployed sensors) but also the increasing demand of new services (by citizens). IoT applications are currently based on multiple architectures, technology standards and seamless software platforms, which have led to a highly fragmented IoT landscape. This fragmentation impacts directly the area of smart cities, which typically comprise several technological silos (i.e. IoT systems that have been developed and deployed independently for smart homes, smart industrial automation, smart transport, and smart buildings etc.).

The operation of IoT applications for Smart Cities will be supported by the introduction of an abstract virtualized digital layer that operate across multiple IoT architectures, platforms (e.g. FI-WARE) and business contexts is required. Smart cities soon will face up the need for an integrated solution(s) (SmartCity-OS) that globally can monitor, visualise and control the uncountable integrated number of operations executed by diverse (and every day increasing) services platforms using the sensor technology deployed in the cities.

The term "Open Data" in the context of Smart Cities generally refers to a public policy that requires public sector agencies and their contractors to release key sets of government data (relating to many public activities of the agency) to the public for any use, or re-use, in an easily accessible manner. In many cases, this policy encourages this data to be freely available and distributable. The value of releasing such data is presumed to lie in the combination of this and other data from various sources. This value can be dramatically increased when the data is discoverable, actionable and available in standard formats for machine readability. The data is then usable by other public agencies, third parties and the general public for new services, and for ever richer insight into the performance of key areas like transport, energy, health and environment. In this context there is a need to ensure that any standards or guidance in this area should not be prescriptive about particular models, but encourage innovation in data re-use [58].

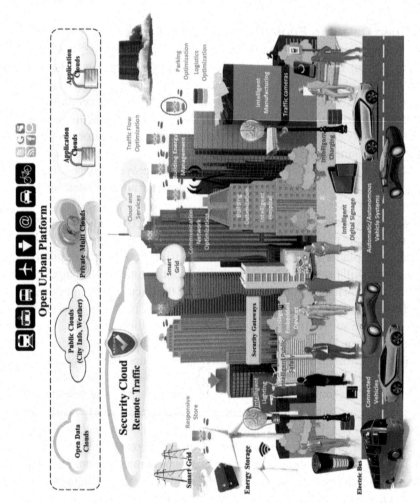

Figure 3.26 Smart City – integration of heterogeneous systems and open data.

The quality of IoT Data and the numerous IoT Data source provisioning are important issues as there is an inherent need to generate semantic-driven business platforms, o address the enabling business-driven IoT ecosystems. These systems have to address functionalities for operating across multiple IoT architectures, platforms and business contexts, to enable a more connected/integrated approach to Smart City applications development.

Smart Cities are becoming one of the biggest fields of application for IoT technologies. Cities are more and more full of devices equipped with sensors, actuators and other appliances providing information that in the past was either impossible or relatively difficult to gather. Their main purpose, among other functionalities, is to gather information about various parameters of importance for management of day-to-day activities in the city as well as for longer term development planning. Examples of such parameters are information about public transport (real-time location, utilization), traffic intensity, environmental data (air quality), occupancy of parking spaces, noise, monitoring of waste bins, energy consumption in public buildings, etc. [66].

Integrated IoT solutions deployed in the cities require addressing interoperability, security, privacy, and trust for all of the suppliers in the

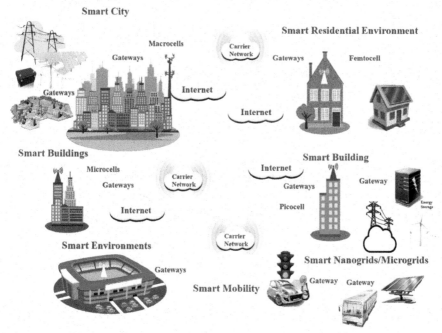

Figure 3.27   Smart City communication technologies landscape.

ecosystem also have policies and safeguards that align to those of the citizens.

The research priorities need to focus on common IoT architecture approaches, IoT data modelling and schema representations, intra-domain and CPS extensions that allows more robustness and extensible IoT platforms with embedded software and applications enabling heterogeneous systems to interact (systems of systems integration) across various verticals in the city.

3.3.8.2 Citizen Centric Smart Cities IoT Applications and Deployments

Public city environments are complex and large. The only possibility to address these largescale, multi-subsystem projects is in a collaborative, open-innovation context, where effort is required to align interests, shape opinions, develop business models and provide a common, interoperable IoT technology ecosystem. Cities are "used" by people, which play different roles on the city (resident citizens, visitors and tourists, businesses, municipal services employees, etc.). The focus on users and citizens can be orchestrated in various dimensions: problems, awareness, participation, culture and digital transformation [66].

In this context, there are numerous important research challenges for smarty city IoT applications:

- Design and implementation of modular architectures enabling easy ways to interface with already existing infrastructures by using standards, protocol wrappers or other innovative means.
- Overcoming traditional silo based organization of the cities, with each utility responsible for their own closed world. Although not technological, this is one of the main barriers.
- Creating algorithms and schemes to describe information created by sensors in different applications to enable useful exchange of information between different city services.
- Mechanisms for cost efficient deployment and even more important maintenance of such installations, including energy scavenging.
- Ensuring reliable readings from a plethora of sensors and efficient calibration of a large number of sensors deployed everywhere from lampposts to waste bins.
- Increasing the intelligence and flexibility on end devices to support them to take autonomous decisions, decreasing resource overloads such as bandwidth and improving their management.

- Provide interoperability solutions that allows that interoperability can be achieved at different levels with the goal of reaching fully interoperability at data level for IoT platforms that operate inside the city and allows the replicability of solutions among cities.
- Design and development of unified APIs for accessing data independently of the protocols, APIs and models supported in the underlying IoT platform in a machine readable way.
- Algorithms for analysis and processing of data acquired in the city and making "sense" out of it.
- IoT large-scale deployment and integration.

3.3.9 Smart Farming and Food Security

Food and fresh water are the most important natural resources in the world. Farming is a major economic activity in Europe [70], with about 12 million farms in the EU-28 in 2010, 40% of the land area and 25 million people dedicated to farming activities. In a European context with its population increasing, achieving higher efficiency in food production is a top priority.

Sustainable farming, producing more with less and with a smaller environmental footprint, is an unstoppable trend that demands new technologies. ICT technologies, and IoT in particular, will be crucial elements for meeting the challenges of tomorrow's sustainable farming, supporting the implementation of smart/precision farming techniques aimed at improving the processes of food production. Indeed, a lot of ICT research and innovation in farming is happening nowadays around precision farming, although the benefits of the application of ICT technologies encompass the whole agri-food value chain as presented in Figure 3.28: food processing, food logistics, wholesale/retail, and finally the consumers.

One crucial aspect that cannot be overlooked, and which is transversal to the whole agri-food value chain, is food safety and traceability: the mechanisms to ensure and monitor those food products are healthy and safe, at their highest possible quality specifications, throughout their whole lifecycle, from farm to fork. Again, food safety can greatly benefit from the application of IoT technologies.

Farming 4.0, or IoT-based innovations applied to farming, has the potential to boost rural areas and EU economy. The AIOTI WG06 Recommendations Report [64], recently published, highlights the benefits that the application of IoT technologies can bring into the agri-food sector, along with the numerous

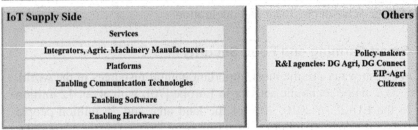

Figure 3.28   Smart farming and food security stakeholders + agri-food value chain.

challenges that must be overcome to unleash their full potential in large scale implementations.

Final IoT-based applications or solutions are enabled by the combination of a number of technology building blocks or layers. Each of those layers faces particular R&I challenges.

IoT applications in the farming sector are dependent on a number of enabling technologies covering hardware (i.e. smart devices that may embed sensors, actuators, communication gateways and other appliances), software (which, embedded in the device, provides it with intelligence, autonomous decision-making, etc.), network/cloud/communication technologies (including the need of reliable, possibly broadband, data coverage in rural or remote areas, and the growing trend of softwarisation/de-hardwarisation and localisation of networks), and services for providing the functionalities needed by the sector. In addition interoperability, standardisation and data management (considering the value and the sensitivity of data generated at farms and other parts of the food chain, but also the added value that comes from data aggregation) are key R&I drivers that are applicable to all technology layers.

A report on smart farming [53] defines seven applications:

- Fleet management – tracking of farm vehicles
- Arable farming, large and small field farming

- Livestock monitoring
- Indoor farming – greenhouses and stables
- Fish farming
- Forestry
- Storage monitoring – water tanks, fuel tanks

Smart farming will allow farmers and growers to improve productivity and reduce waste, ranging from the quantity of fertiliser used to the number of journeys made by farm vehicles. The complexity of smart farming is also reflected into the ecosystem of players. They can be classified in the following way:

- Technology providers – these include providers of wireless connectivity, sensors, M2M solutions, decision support systems at the back office, big data analytical systems, geo-mapping applications, smartphone apps
- Providers of agricultural equipment and machinery (combines, tractors, robots), farm buildings, as well as providers of specialist products (e.g. seeds, feeds) and expertise in crop management and animal husbandry
- Customers: farmers, farming associations and cooperatives
- Influencers – those that set prices, influence the market into which farmers and growers sell their products.

The range of stakeholders in agriculture is broad, ranging from big business, finance, engineering, chemical companies, food retailers to industry associations and groupings through small suppliers of expertise in all the specialist areas of farming.

The end users of precision farming solutions include not only the growers but also farm managers, users of back office IT systems. Not to be forgotten is the role of the veterinary in understanding animal health. Also to be considered are farmers co-operatives, which can help smaller farmers with advice and funding.

The following table provides an overview of the most relevant challenges across the technology layers.

Table 3.1 Technological challenges for IoT applications in the farming sector

Development	2016–2020	Beyond 2020
Enabling hardware	• Improve the ratio computational power-to-energy consumption of devices, possibly combined with energy harvesting or local renewable generation.	• Implementation of more efficient hardware cryptographic primitives embedded in hardware devices

(Continued)

Table 3.1 Continued

Development	2016–2020	Beyond 2020
	• Increase hardware robustness: longer lifetime and calibration cycles • Development of cost-effective near-field communication technologies suitable for massive use in food products	
Enabling software	• Development of flexible real-time and embedded micro operating systems • Self-configurable, remotely attestable devices • Large-scale device management and orchestration software and middleware, including SW	• Self-configurable, remotely attestable devices
Enabling network, cloud, communication technologies	• SDN/NFV for telcos targeted to smart agriculture applications • Edge analytics to promote local data circulation • Definition and application of protocols with bounded message delivery times (for real-time applications) • Federated/orchestrated hybrid clouds and transition to communal equipment/infrastructure • Level playing field facilitating competition among service providers • Increase the range of communication and reliability of deployed devices • Adapt communications architecture for supporting low individual device throughput and high aggregated network throughput (i.e. few short messages from each device, but a high amount of individual data sources) • Automatic deployment (no need for configuration of the communications)	• SDN/NFV for telcos targeted to smart agriculture applications • Distributed communication architectures (e.g. Edge Computing) to treat smart farming as critical industries in terms of time latencies
Service layer	• Data analytics and predictive modelling for decision-support systems	• Data analytics and predictive modelling for decision-support systems

	• High accuracy (indoor and outdoor) positioning and mapping solutions cost-effective enough for smaller farms to adopted precision farming • Farm management systems and precision farming solutions easily adaptable to holdings of different sizes • Service providing infrastructure for 3rd parties allowing the integration of external service providers that use internal data (for example, a company that provides irrigation optimization analysis) • Farm management systems satisfying energy efficiency objectives, related to cultivation and farm management processes • User interfaces with high usability and low learning curve • Stimulate innovation in targeting cross-sectorial IoT applications such as smart energy management for farms, smart nutrition management for end-consumers	• Farm management systems satisfying energy efficiency objectives, related to cultivation and farm management processes
Interoperability and standardisation	• Specification and implementation of protocols for agricultural machinery information exchange, including fleet management • Development of open reference vocabularies, formats and protocols for data storage and exchange allowing flexible interaction between arbitrary actors across the food chain • Specification of universal identification standards and technologies inter-linking among different addressing techniques, to make sure those different parts in food traceability scenarios can be properly referred to and logically interrelated.	• Development of open reference vocabularies, formats and protocols for data storage and exchange allowing flexible interaction between arbitrary actors across the food chain
Data management and protection	• Digital Rights Management in the farming domain, including scenarios of data aggregation and data sharing	• Trusted data: integrity and authenticity of the data generated/stored.

(Continued)

Table 3.1   Continued

Development	2016–2020	Beyond 2020
	• Trusted data: integrity and authenticity of the data generated/stored. The origin of the product, the processing stages it passed through and other sensitive information must be known. Guarantee the trustworthiness of the source is a crucial requirement. • Low cost authentication mechanisms for devices/machines • Access control policies and access control mechanisms for individual users and individual pieces of information • Develop hybrid cloud storage and interaction models which unite the universal data availability of cloud solutions with the individual, local control of data owners and the resilience against disruptive crisis provided by de-centralized island networks and individualized peer-to-peer communication	The origin of the product, the processing stages it passed through and other sensitive information must be known. Guarantee the trustworthiness of the source is a crucial requirement. • Low cost authentication mechanisms for devices/machines

3.3.9.1 Business Models and Innovation Ecosystems

The deployment and adoption of IoT technologies and applications in the farming sector need to address the different challenges and opportunities created by the new business models introduced. A number of issues that have to be considered are presented below:

- Provide evidence of the sustainability of the IoT-based business, both for the supply (ICT) and demand (agri-food) sides. From the point of view of the users, the quantifiable benefit and profitability must compensate for the cost of the IoT solutions.
- A challenge, and at the same time an opportunity, is the possibility of devising new, disruptive business models. Some traditional companies, for instance, are already shifting their business to data-driven models.
- Stimulate and empower the role of consumers as key element/ beneficiaries of the IoT-enabled food supply chain

- Build trust around the smart farming technology made in the EU (for example through a IoT trust label)
- Analyse the important role of farm advisory services in the context of data-driven farming
- Foster the creation of digital farming innovation hubs, not only in EU, but at regional/national level, to accelerate innovation and adoption, facilitate the early exchange of best practice.

3.3.9.2 Societal Aspects

The complexity of smart farming and the proliferation of IoT technologies provided by various stakeholders or ecosystems requires considering the following social aspects when addressing the implementation and deployments:

- Identify the lack of digital skills preventing the adoption of digital agriculture in some EU regions, and take corrective action involving the necessary stakeholders (cooperatives, regional administrations) in order to prevent a digital divide in EU's agriculture.
- Provide evidence of the positive impact of the digitisation of farming in the EU's rural economy. Analyse new potential relationships between the rural and urban economies.
- Stimulate and empower the role of consumers as key element/ beneficiaries of the IoT-enabled food supply chain
- Promote transparency of the food production process and encourage data sharing by farmers along agri-food value chain
- Take action to ensure that the benefits of IoT reach all types of farms, especially smaller and family-owned holdings, which constitute the vast majority in Europe, and thus are of utmost socioeconomic importance

3.3.9.3 Coordination among Different DGs, Programmes and Member States

Although H2020 can help by providing a spearhead or lighthouse in the form of a Large Scale Pilot, a large amount of IoT take-up in the farming sector will be happening in parallel under national or regional initiatives (and thus in a smaller, more fragmented scale). Much of this technology take-up can or will be facilitated by public investments of Structural Funds or other funding sources managed at a national or even regional level, such as EAFRD (European Agricultural Fund for Rural Development, implementing the Common Agricultural Policy 2014–2020, CAP) and ERDF (European

Regional Development Fund), the latter in regions with Smart Specialization Strategies.

The active coordination of the different Administrations involved (EU, national and regional) towards streamlining efforts and generating operational efficiencies can only contribute to maximize the chances of having a vibrant smart farming ecosystem in the EU benefitting users and providers alike, as well as consumers and the European society and economy.

National, regional and cities' Public Administrations can play an important role as either users, infrastructure managers, procurers, initial demand facilitators or subsidizers. In this sense, it is important to consider the aggregation of national and regional initiatives related to IoT for pre-commercial procurement, deployment, coordination of R&I programmes, etc, and the exchange of best-practices among leading Member States/Regions and followers/laggards.

Following the IoT cross-cutting actions implemented in the H2020 Work Programme 2016–17, further collaboration in the design of new work programmes is highly desirable among DG CONNECT, DG AGRI, DG RESEARCH, DG MARE (to include aquaculture in the future actions), as well as DG ENER and DG Health and Food Safety.

3.3.9.4 Policy and Regulations

In the context of the DSM, the barriers blocking widespread deployment of IoT-based innovations in farming (including interoperability, connectivity, and security) must be lowered. The agrifood sector should be no exception in benefitting from the more agile digital economy. In this vein, policy makers could benefit from a sound analysis of major threats: data management and trust (ownership, rules for access, security, and, where applicable, privacy), connectivity and internet access in rural areas, cost of high accuracy positioning services, and digital literacy and skills, among others.

In the context of the EU Common Agricultural Policy (CAP), whose primary objective nowadays is market-oriented sustainable food production, mechanisms could be designed to supporting the adoption of digital technologies in farming uniformly across the EU.

Regulations regarding traceability and labelling should be addressed to facilitate adoption of new IoT solutions for traceability at EU-wide level.

3.4 IoT and Related Future Internet Technologies

3.4.1 Cloud Computing

The Cloud computing definition provided by the National Institute of Standard and Technologies (NIST) covers the main features of the technology. The definition states that the cloud computing is a model for enabling ubiquitous, convenient, on-demand network access to a shared pool of configurable computing resources (e.g., networks, servers, storage, applications, and services) that can be rapidly provisioned and released with minimal management effort or service provider interaction [25].

Figure 3.29 summarises the main aspects of cloud, characteristics, the layered architecture and the standard service models. In the following, we describe a few important aspects of Cloud. The architecture of Cloud can be split into several layers: datacentre (hardware), infrastructure, platform, and application. Each of them can be seen as a service for the layer above and as a consumer for the layer below. Cloud services can be grouped in three main categories: Software as a Service (SaaS), Platform as a Service (PaaS), and Infrastructure as a Service (IaaS). SaaS refers to the provisioning of applications running on Cloud environments. Applications are typically accessible through a thin client or a web browser. PaaS refers to platform-layer

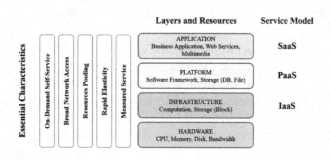

Figure 3.29 Cloud paradigm [24].

resources (e.g., operating system support, software development frameworks, etc.). IaaS refers to providing processing, storage, and network resources, allowing the consumer to control the operating system, storage and applications [24]. IoT can benefit from the capabilities and resources of cloud to compensate its technological constraints (e.g., storage, processing, communication, etc.). Cloud can offer an effective solution for IoT service management and composition as well as for implementing applications and services that exploit the things or the data generated by the things. Cloud can benefit from IoT by extending its usage to deal with real world things in a more distributed and dynamic manner, and for delivering new services in a large number of real life scenarios.

Cloud computing provide a unique opportunity to unify the real, digital and the virtual worlds. IoT enables the building of very large infrastructures that facilitate the information-driven real-time integration of the physical world, sensing/actuating, processing, analytics, with the digital, cyber and virtual worlds on a global scale.

3.4.2 Edge Computing

Virtualisation of objects will push for the convergence of cloud computing and IoT will enable unprecedented opportunities in the IoT services arena [80]. The central idea is that IoT's biggest transformation will be in shifting power in a network from the center to the edge. Rather than devices and users communicating through central hubs – mainframes or cloud based management servers, IoT will allow devices to communicate directly with each other, which is the implementation of the "democratic" vision of a decentralized Internet [82].

The IoT layered architecture include the edge intelligence into the edge computing/processing where all the data capture, processing is done at the device level among all the physical sensor/actuators/devices that include controllers based on microprocessors/microcontrollers to compute/process and wireless modules to communicate. The intelligence at the edge supports devices to use their data sharing and decision-making capabilities to interact and cooperate in order to process the data at the edge, filter it and select/prioritize what is important.

This intelligent processing at the edge select the "smart data" that is transferred to the central data stores for further processing in the cloud. This allows including the Edge Cloud for processing data and addressing the challenges of response-time, reliability and security. For real time fast

Figure 3.30 Evolution of IoT from centralised networks to distributed cloud [82].

processes, the sensor/actuator edge devices could generate data much faster than the cloud-based apps can process it.

The use of intelligent edge devices require to reduce the amount of data sent to the cloud through quality filtering and aggregation and the integration of more functions into intelligent devices and gateways closer to the edge reduces latency. By moving the intelligence to the edge, the local devices can generate value when there are challenges related to transferring data to the cloud. This will allow as well for protocol consolidation by controlling the various ways devices can communicate with each other.

As part of this convergence, IoT applications (such as sensor-based services) will be delivered on-demand through a cloud environment [81]. This extends beyond the need to virtualize sensor data stores in a scalable fashion. It asks for virtualization of Internet-connected objects and their ability to become orchestrated into on-demand services (such as Sensing-as-a-Service).

Computing at the edge of the mobile network defines the IoT-enabled customer experiences and require a resilient and robust underlying network infrastructures to drive business success. IoT assets and devices are connected via mobile infrastructure, and cloud services are provided to IoT platforms to deliver real-time and context-based services.

Data transmission costs and the latency limitations of mobile connectivity pose challenges to many IoT applications that rely on cloud computing. Mobile edge computing will enable businesses to deliver real-time and context-based mobile moments to users of IoT solutions, while managing the cost base for mobile infrastructure. A number of challenges listed below have to be addressed when considering edge-computing implementation [83]:

- Cloud computing and IoT applications are closely connected and improve IoT experiences. IoT applications gain functionality through cloud services, which in turn open access to third-party expertise and up-to-date information.
- Mobile connectivity can create challenges for cloud-enabled IoT environments. Latency affects user experiences, so poor mobile connectivity can limit cloud-computing deployments in the IoT context.
- Mobile edge computing provides real-time network and context information, including location, while giving application developers and business leaders access to cloud computing capabilities and a cloud service environment that's closer to their actual users.
- Mobile edge computing is an important network infrastructure component for block chain. The continuous replication of "blocks" via devices

on this distributed data centre poses a tremendous technological challenge. Mobile edge computing reveals one opportunity to address this challenge.

Edge computing refers to data processing power at the edge of a network and in industrial IoT applications (i.e. power production, smart traffic lights, manufacturing, etc.) the edge networked devices capture data and process date close to the source of performing "edge analytics" on the data. Edge computing complement cloud computing, since an analytic model or rules are created in the cloud then pushed out to edge devices. Edge computing is closely related to fog computing, that entails data processing from the edge of the network to the cloud.

For the future IoT applications it is expected that more of the network intelligence to reside closer to the source. This will push for the rise of Edge Cloud/Fog, Mobile Edge computing architectures, as most data will be too noisy or latency-sensitive or expensive to be transfer to the cloud.

The previous IERC SRIAs have identified the importance of interoperability semantic technologies towards discovering devices, as well as towards achieving semantic interoperability.

3.5 Networks and Communication

The IERC SRIA intends to lay the foundations for the IoT to be developed by research through to the end of this decade and for subsequent innovations to be realised even after this research period. Within this timeframe, the number of connected devices, their features, their distribution and implied communication requirements will develop, as will the communication infrastructure and it is predicted that low-power short-range networks will dominate wireless IoT connectivity through 2025, far outnumbering connections using wide-area IoT networks [21]. IoT technologies are extending the known business models and leading to the proliferation of different ones as companies push beyond the data, analytics and intelligence boundaries, while everything will change significantly. IoT devices will be contributing to and strongly driving this development.

Changes will first be embedded in given communication standards and networks and subsequently in the communication and network structures defined by these standards.

Further developments of networks and communication technologies are required by the emergence of the Tactile Internet, in which ultra-responsive and ultra-reliable network connectivity will enable it to deliver physical haptic experiences remotely for different IoT applications. The Tactile Internet will add a new dimension to human-machine interaction through building real-time interactive systems. The combination of Tactile Internet and IoT applications will enable haptic communications at the edge and in the inter-action between humans and machines, infrastructure and environment by providing the medium for transporting touch and actuation in real-time i.e., the ability of haptic control through the Internet, in addition to no haptic control and data.

3.5.1 Network Technology

The development in cloud and mobile edge computing requires network strategies for fifth evolution of mobile the 5G, which represents clearly a convergence of network access technologies. The architecture of such network has to integrate the needs for IoT applications and to offer seamless integration and optimise the access to Cloud or mobile edge computing resources. IoT is estimated that will connect 30 billion devices. All these devices are connecting humans, things, information and content, which is changing the performance characteristics of the network. Low latency is becoming crucial (connected vehicles or industrial equipment must react in ms), there is a need to extend network coverage even in non-urban areas, a better indoor coverage is required, ultra-low power as many of the devices will be battery operated is needed and a much higher reliability and robustness is requested.

5G networks will deliver 1,000 to 5,000 times more capacity than 3G and 4G networks today and will be made up of cells that support peak rates of between 10 and 100 Gbps. They need to be ultra-low latency, meaning it will take data 1–10 milliseconds to get from one designated point to another, compared to 40–60 milliseconds today. Another goal is to separate communications infrastructure and allow mobile users to move seamlessly between 5G, 4G, and WiFi, which will be fully integrated with the cellular network. To support the increasing data rates and number of connected devices in urban environments, mobile networks are increasingly dense and heterogeneous in cell-size and radio access technologies (multi-RATs).

Applications making use of cloud computing, and those using edge computing will have to co-exist and will have to securely share data. The right balance needs to be found between cloud/mobile edge computing to optimize overall network traffic and optimize the latency. Facilitating optimal use of both mobile edge and cloud computing, while bringing the computing processing capabilities to the end user. Local gateways can be involved in this optimization to maximize utility, reliability, and privacy and minimize latency and energy expenditures of the entire networks.

Future networks have to address the interference between the different cells and radiations and develop new management models control roaming, while exploiting the co-existence of the different cells and radio access technologies. New management protocols controlling the user assignment to cells and technology will have to be deployed in the mobile core network for a better efficiency in accessing the network resource. Satellite communications need to be considered as a potential radio access technology, especially in remote areas. With the emerging of safety applications, minimizing the latency and the various protocol translation will benefit to the end-to-end latency. Densification of the mobile network strongly challenges the connection with the core network. Future networks should however implement cloud utilization mechanisms to maximize the efficiency in terms of latency, security, energy efficiency and accessibility.

In this context, there is a need for higher network flexibility combining Cloud technologies with Software Defined Networks (SDN) and Network Functions Virtualisation (NFV), that will enable network flexibility to integrate new applications and to configure network resources adequately (sharing computing resources, split data traffic, security rules, QoS parameters, mobility, etc.).

The evolution and pervasiveness of present communication technologies has the potential to grow to unprecedented levels in the near future by including the world of things into the developing IoT. Network users will be humans, machines, things and groups of them.

3.5.2 Communication Technology

The growth in mobile device market is pushing the deployment of IoT applications where these mobile devices (smart phones, tablets, etc. are seen as gateways for wireless sensors and actuators.

Communications technologies for the Future Internet and the IoT will have to avoid such bottlenecks by construction not only for a given status of development, but also for the whole path to fully developed and still growing nets.

The inherent trend to higher complexity of solutions on all levels will be seriously questioned – at least with regard to minimum energy IoT devices and services.

Their communication with the access edges of the IoT network shall be optimized cross domain with their implementation space and it shall be compatible with the correctness of the construction approach.

These trends require the extension of the spectrum in to the 10–100 GHz and unlicensed band and technologies like WiGig or 802.11ad that are mature enough for massive deployment, can be used for cell backhaul, point-to-point or point-to-multipoint communication. The use of advanced multi-/massive-MIMO technologies have the capability to address both coverage and bandwidth increase, while contributing to optimize the usage of the network resources adequately to real need.

The IoT applications will embed the devices in various forms of communication models that will coexist in heterogeneous environments. The models will range from device to device, device to cloud and device to gateway communications that will bring various requirements to the development of electronic components and systems for IoT applications. The first approach considers the case of devices that directly connect and communicate between each another (i.e. using Bluetooth, Z-Wave, ZigBee, etc.) not necessarily using an intermediary application server to establish direct device-to-device communications. The second approach considers that the IoT device connect (i.e. using wired Ethernet or Wi-Fi connections) directly to Internet cloud/fog service of various service providers to exchange data and control message traffic. The third approach, the IoT devices connect to an application layer gateway running an application software operating on the gateway device, providing the "bridge" between the device and the cloud service while providing security, data protocol translation and other functionalities.

The deployment of billions of devices requires network agnostic solutions that integrate mobile, narrow band IoT (NB IoT), LPWA networks, (LoRA, Sigfox, Weightless, etc), and high speed wireless networks (Wi-Fi), particularly for applications spanning multiple jurisdictions.

LPWA networks have several features that make them particularly attractive for IoT devices and applications that require low mobility and low levels of data transfer:

- Low power consumption that enable devices to last up to 10 years on a single charge
- Optimised data transfer that supports small, intermittent blocks of data

Table 3.2 LPWA network protocols

Name of Standard	Weightless -W	Weightless -N	Weightless -P	SigFox	LoRaWAN	LTE-Cat M	IEEE P802.11ah (LP WiFi)	Dash7 Alliance Protocol 1.0	Ingenu RPMA	nWave
Frequency Band	TV white-space (400–800 MHz)	Sub-GHZ ISM	Sub-GHZ ISM	868 MHz/902 MHz ISM	433/868/780/915 MHz ISM	Cellular	License-exempt bands below 1 GHz, excluding the TV White Spaces	433, 868, 915 MHz ISM/SRD	2.4 GHz ISM	Sub-GHz ISM
Channel Width	5 MHz	Ultra narrow band (200 Hz)	12.5 kHz	Ultra narrow band	EU: 8×125 kHz, US 64×125 kHz/8×125 kHz, Modulation: Chirp Spread Spectrum	1.4 MHz	1/2/4/8/16 MHz	25 KHz or 200 KHz	1 MHz (40 channels available)	Ultra narrow band
Range	5 km (urban)	3 km (urban)	2 km (urban)	30–50 km (rural), 3–10 km (urban), 1000 km LoS	2–5 k (urban), 15 k (rural)	2.5–5 km	Up to 1 km (outdoor)	0–5 km	>500 km LoS	10 km (urban), 20–30 km (rural)

(Continued)

Table 3.2 Continued

Name of Standard	Weightless			SigFox	LoRaWAN	LTE-Cat M	IEEE P802.11ah (LP WiFi)	Dash7 Alliance Protocol 1.0	Ingenu RPMA	nWave
	-W	-N	-P							
End Node Transmit Power	17 dBm	17 dBm	17 dBm	10 µW to 100 mW	EU: < +14dBm, US: < +27 dBm	100 mW	Dependent on Regional Regulations (from 1 mW to 1 W)	Depending on FCC/ETSI regulations	to 20 dBm	25–100 mW
Packet Size	10 byte min.	Up to 20 bytes	10 byte min.	12 bytes	Defined by user	~100–1000 bytes typical	Up to 7,991 Bytes (w/o aggregation), up to 65,535 Bytes (with aggregation)	256 bytes max/packet	Flexible (6 bytes to 10 kbytes)	12 byte header, 2–20 byte payload

Uplink Data Rate	1 kbps to 10 Mbps	100 bps	200 bps to 100 kbps	100 bps to 140 messages/day	EU: 300 bps to 50 kbps, US: 900–100 kbps	~200 kbps	150 Kbps ~346.666 Mbps	9.6 kb/s, 55.55 kbps or 166.667 kb/s	AP aggregates to 624 kbps per Sector (Assumes 8 channel Access Point)	100 bps
Downlink Data Rate	1 kbps to 10 Mbps	No downlink	200 bps to 100 kbps	Max. 4 messages of 8 bytes/day	EU: 300 bps to 50 kbps, US: 900–100 kbps	~200 kbps	150 Kbps ~346.666 Mbps	9.6 kb/s, 55.55 kbps or 166.667 kb/s	AP aggregates to 156 kbps per Sector (Assumes 8 channel Access Point)	–
Devices per Access Point	Unlimited	Unlimited	Unlimited	1 M	Uplink: > 1, Downlink: < 100 k	20k+	8191	NA (connectionless communication)	Up to 384,000 per sector	1 M

(Continued)

Table 3.2   Continued

| Name of Standard | Weightless | | | SigFox | LoRaWAN | LTE-Cat M | IEEE P802.11ah (LP WiFi) | Dash7 Alliance Protocol 1.0 | Ingenu RPMA | nWave |
	-W	-N	-P							
Topology	Star	Star	Star	Star	Star on Star	Star	Star, Tree	Node-to-node, Star, Tree	Typically Star. Tree supported with an RPMA extender	Star
End node roaming allowed	Yes	Yes	Yes	Yes	Yes	Yes	Allowed by IEEE 802.11 amendments (e.g., IEEE 802.11r)	Yes	Yes	Yes
Governing Body	Weightless SIG			Sigfox	LoRa Alliance	3GPP	IEEE 802.11 working group	Dash7 Alliance	Ingenu (OnRamp)	Weightless SIG

- Low device unit cost
- Few base stations required to provide coverage
- Easy installation of the network
- Dedicated network authentication
- Optimised for low throughput, long or short distance
- Sufficient indoor penetration and coverage

These different types of networks are needed to address IoT product, services and techniques to improve the Grade of Service (GoS), Quality of Service and Quality of Experience (QoE) for the end users. Customization-based solutions, are addressing industrial IoT while moving to a managed wide-area communications system and, ecosystem collaboration.

Intelligent gateways will be needed at lower cost to simplify the infrastructure complexity for end consumers, enterprises, and industrial environments. Multi-functional, multi-protocol, processing gateways are likely to be deployed for IoT devices and combined with Internet protocols and different communication protocols.

These different approaches show that device interoperability and open standards are key considerations in the design and development of internetworked IoT systems.

Ensuring the security, reliability, resilience, and stability of Internet applications and services is critical to promoting the concept of trusted IoT based on the features and security provided of the devices at various levels of the digital value chain.

3.6 IoT Standardisation

In recent publications mapping emerging technologies to their Hype Cycle, Gartner positions the IoT at the top of the "Peak of Inflated Expectations" [14].

The assessment is widely shared and is reflected by significant IoT related activities in companies of all sizes, in industry standards groups, consortia, alliances and in the press and media. Many observers also remark on the number of technologies, alliances and consortia across the IoT landscape and agree that a consolidation is imminent. These expectations broadly align with the lifecycle phases that Gartner's model predicts for IoT. Gartner's view is that IoT will reach the "Plateau of Productivity" in 5–10 years – somewhere around 2020–2025. On that basis, they anticipate that the period 2015–2019 will see a consolidation phase with a corresponding reduction in hype, a period of intense development of standards, and a transition into a period of real product development.

Table 3.3 Standardisation key challenges addressed by AIOTI

Domain	Activities
Architecture	• Guidelines and recommendations, which contribute to the consolidation of architectural frameworks, reference architectures, and architectural styles in the IoT space.
Semantic Interoperability	• Guidelines and recommendations, which contribute to the consolidation of semantic interoperability approaches in the IoT space.
Privacy	• Guidelines and recommendations regarding personal data and personal data protection to the various categories of stakeholders in the IoT space.

Standardisation will play a key role in the consolidation of IoT landscape; since many of the benefits of IoT will occur based on widespread adoption, the development of global standards is pivotal to ensuring economies of scale and impact.

The standardisation priorities for AIOTI WG03 [61] will be a focus of European engagement and steering in the standardization process. In collaboration with other AIOTI working groups, the focus will be to:

- Maintain a view on the landscape of IoT standards-relevant activities being driven by SDOs, Consortia, Alliances and OSS projects.
- Provide a forum for analysis, discussion and alignment of strategic, cross-domain, technical themes and shared concerns across landscape activities
- Develop recommendations and guidelines addressing those concerns
- Engage the IoT community in disseminating and promoting the results and steering emerging standards

In collaboration with ST505, AIOTI WG03 will build an understanding of SDOs, Alliances, and Consortia; their respective specifications, technologies, and spheres of influence; and the breadth, depth and sustainability of any Open Source Software, which has established a usage profile.

The outputs of the landscape work will drive the WG03 program. Analysis of gaps, divergences, common concerns, and major players will inform the agenda of challenges to be addressed, guidelines and recommendations to be developed and groups to be engaged with.

The following table provides the three key challenges the workgroup is currently responding to.

AIOTI WG03 will support the implementation of the goals set by the EC [16] and promote the use of open standards through actions that: (1) support the entire value chain, (2) apply within IoT domains and cross-IoT domains

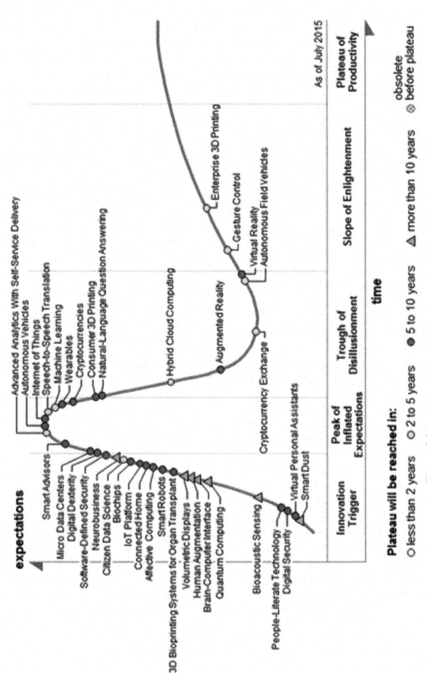

Figure 3.31 Hype Cycle of emerging technologies [14].

Table 3.4   Standardisation challenges for IoT

Specific IoT Standardisation Challenges	
2016–2020	Beyond 2020
• Recommendations of reference architectures, both for experimentation and deployments within IoT domains and cross IoT domains	
• Identification of missing (semantic) interoperability standards and technologies within IoT domains and cross IoT domains and recommendations on solving them	
• Recommendations and guidelines on solving protocol and interface gaps needed to support new IoT features within IoT domains and cross IoT domains. Promote the uptake of IoT standards in public procurement to avoid lock-in	• Further work on recommendations and guidelines on solving protocol and interface gaps needed to support new IoT features within IoT domains and cross IoT domains. Promote the uptake of IoT standards in public procurement to avoid lock-in
• Promoting the use and development of Open Reference Vocabularies and Open Application Programming Interfaces to allow for flexible ad-hoc communication and interaction between different actors within IoT domains and cross IoT domains	• Further development and promotion of the use and development of Open Reference Vocabularies and Open Application Programming Interfaces to allow for flexible ad-hoc communication and interaction between different actors within IoT domains and cross IoT domains
• Provide guidelines on how to translate the Digital Rights Management recommendations within IoT domains and cross IoT domains	
• Recommendation of an interoperable IoT numbering space that transcends geographical limits, and an open system for object identification and authentication, which can be applied within IoT domains and cross IoT domains	
• Explore options and recommend guiding principles, including guidelines for the support of developing standards, for trust, privacy and end-to-end security, e.g. through a 'trusted IoT label' that can be applied within IoT domains and cross IoT domains	• Explore options and recommend guiding principles, including guidelines for the support of developing standards, for trust, privacy and end-to-end security, e.g. through a 'trusted IoT label' that can be applied within IoT domains and cross IoT domains

and (3) are integrating multiple technologies. This is done based on streamlined international cooperation, which enables easy and fair access to standard essential patents (SEPs). In order to accomplish this goal several potential challenges can be foreseen, which are presented in the following table.

3.7 IoT Security

Security needs to be designed into IoT solutions from the concept phase and integrated at the hardware level, the firmware level, the software level and the service level. IoT applications need to embed mechanisms to continuously monitor security and stay ahead of the threats posed by interactions with other IoT applications and environments. Trust is based on the ability to maintain the security of the IoT system and the ability to protect application/customer information, and as well as being able to respond to unintended security or privacy breaches. In the IoT it is important to drive security, privacy, data protection and trust across the whole IoT ecosystem and no company can "do it alone" in the IoT space; success will require organizations to partner, value chains to be created and ecosystems to flourish. Yet as IoT users start to bring more players, service providers and third party suppliers into their value chain, tech firms and IoT solutions providers will face increasing pressure to demonstrate their security capabilities [10].

The worlds of IT and operational technology (OT) are converging, and IT leaders must manage their transition to converging, aligning and integrating IT and OT environments [12]. The benefits that come from managing IT and OT convergence, alignment and integration include optimized business processes, enhanced information for better decisions, reduced costs, lower risks and shortened project timelines. IT and OT are converging in numerous important industries, such as healthcare, transportation, defence, energy, aviation, manufacturing, engineering, mining, oil and gas, natural resources and utilities. IT leaders who are impacted by the convergence of IT and OT platforms should consider the value and risk of pursuing alignment between IT and OT, as well as the potential to integrate people, tools and resources used to manage and support both technology areas. A shared set of standards and platforms across IT and OT will reduce costs in many areas of software management, while the reduction in risks that will come from reducing malware intrusion, internal errors and cybersecurity can be enhanced if IT security teams are shared, seconded or combined with OT staff to plan and implement holistic IT-OT security [12].

The evolution of connected devices as nodes to the IoT brings limitless possibilities. As more and more everyday things are connected to the Internet – medical devices, automobiles, homes, etc. – the long-term forecast for the IoT is staggering: by 2020, there will be 212 billion installed things, 30 billion autonomously connected things and approximately three million petabytes of embedded system data, all of which combined are expected to generate nearly $9 trillion in business value. IoT applications fall into three basic categories [11]:

- Mobile or desktop applications that control IoT devices;
- IoT firmware and embedded applications;
- Applications on open IoT platforms (for example, apps built for Apple Watch).

All of these applications need to be protected or they run the risk of undesirable outcomes such as:

- Improper or unsafe operation of IoT devices;
- Theft of confidential data, private user information or application-related intellectual property;
- Fraud and unauthorized access to payment processing channels;
- Damage to companies brand image and deterioration of customer, prospect and partner trust.

In the case of IoT, applications can be attacked in many ways, often involving apps that first obtain access to the IoT application, then start monitoring, controlling, and tampering with the device.

A holistic approach that involves the device, data, network and application layers is required and the following chart summarizes key IoT security components that must be considered [11]:

The following policy recommendations on net neutrality and IoT, given the current relevance of net neutrality to the European policy debate, following agreement of the Telecoms Single Market legislative package are given in [62] and summarised below:

- Embed "safe and secure software" design and development methodologies across all levels of device/application design and development and implement security into that life cycle at the same time.
- Design, deliver and operate adaptive and dynamic end-to-end security over heterogeneous infrastructures integrating IoT, networks and cloud infrastructures. It is recommended to use underlying standardised OS and hardware security features where architecture permits. The deployment

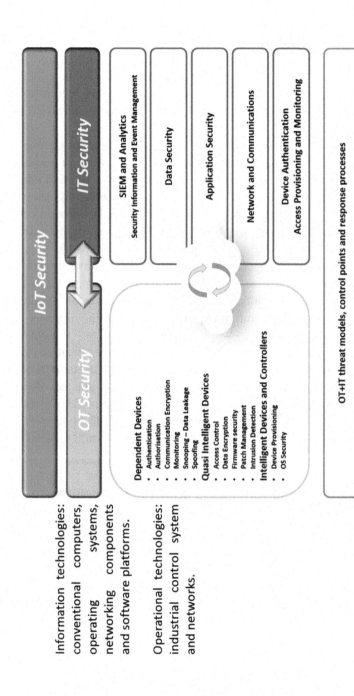

Figure 3.32 IoT security challenges for IT and OT technologies (Adapted from [11]).

should not be specific or propose a modification of existing OS and hardware already integrated by IoT.

- Develop best practices confirming minimum requirements for provision of secure, encrypted and integrity-protected channel, mutual authentication processes between devices and measures securing that only authorised agents can change settings on communication and functionality.
- Develop a "New Identity for Things" – To date, Identity and Access Management (IAM) processes and infrastructures have been primarily focused on managing the identities of people. IAM processes and infrastructure must now be re-envisioned to encompass the amazing variety of the virtualized infrastructure components. For example, authentication and authorization functions will be expanded and enhanced to address people, software and devices as a single converged framework.
- Develop a Common Authentication architecture – by investigation of a Secure Identity and Trusted Authentication mechanism, for example one which takes into account different authentication standards and will provide a single-sign-on solution for IoT applications moving between different systems.
- Certification – the certification framework and self-certification solutions for IoT applications have not been developed yet. The challenge will be to have generic and common framework, while developing business specific provisions. This framework should provide evaluation assurance levels similar to the Common Criteria for Information Technology Security Evaluation (ISO/IEC 15408), which should serve as the reference.

3.7.1 IoT Security Framework based on Artificial Intelligence Concepts

Large-scale applications and services based on the IoT are increasingly vulnerable to disruption from attack or information theft. Vulnerability in general terms is defined as the opportunity for a threat to cause loss. A threat is any potential danger to a resource, originating from anything or anyone that has the potential to cause a threat. Clearly, specific and more intelligent security solutions are required to cope with these issues, which if not addressed may become barriers for the IoT deployment on a broad scale.

Swarm intelligence (SI) is such a technological area, which can inspire the design of new IoT security solutions. A subfield of artificial intelligence,

SI studies the emergent collective intelligence of groups of agents based on social behaviour that can be observed in nature, such as ant colonies, flocks of birds, fish schools and bee hives, where a number of individuals with limited capabilities are able to produce intelligent solutions for complex problems. Vulnerability and reaction to threats seem to be a common thread and IoT can take inspiration from ant colonies, flocks of birds, fish schools and bee hives on how to react to threats.

IoT objects have more capabilities than the above examples; in fact, the trend is towards distributed models, meaning that objects are becoming more intelligent, capable of making their own authentication, authorization and other trust management decisions. Nevertheless, by embracing principles of swarm intelligence, IoT systems can react more effectively to threats. Clearly, a group of IoT objects has more abilities and resources to process large amounts of information in real time in order to prevent, detect and react to perceived or real threats, as well as make decisions based on the acquired information.

The idea is not to make the IoT objects mobile in order to physically group objects when threats occur but to augment the intelligence internalized in each object, with new kind of intelligence that allow the individual resources and intelligence in objects to group. Not all objects need to group at all times. Objects can group around one object identified as a point of attack or around a path of objects.

Clustering is therefore an important area and has been applied in many domains, such as spatial data analysis, image processing, marketing and pattern recognition, etc. For example, ant-based clustering is a type of clustering algorithm that imitates the behaviour of ants, with a perfect social organization where each type of individual specializes in a specific activity within the colony.

In IoT security, the purpose of clustering is to cluster IoT objects into groups according to some predefined rules addressing the issues inherent in detecting and dealing with threats.

The essence of this concept can be best illustrated by the following rules of separation, alignment, cohesion of the first multi-agent algorithm developed by Craig Reynolds in 1986 simulating swarm behaviour.

- Separation: going away from other agents. In the IoT context, this rule would become preserving the distributed nature of the IoT system in the absence of threats, so that individual resources can be focused on the functions to be performed by each object. Unnecessary clustering would consume resources and would even expose intelligence crowding to attack.

- Cohesion: going to the centre of the surrounding agents. In the IoT context, this rule would become steering resources and intelligence towards one or several points of attack.
- Alignment: heading towards the same direction of other agents. In the IoT context, this rule would become steering along a path of attack.

Complex behaviour can be programmed as rules, based on self-organization. The basic concept is to define rules and constraints and let the IoT system self-organize in the presence of threats. The self-organization properties may help security architects and other professionals to discover new security solutions.

3.7.2 Self-protecting, Self-optimizing and Self-healing IoT Concepts

Self-protecting capability features opens up the possibility for IoT to be used in systems that need to protect themselves from malicious attacks, because security, privacy and data protection are at stake.

IoT may offer other capabilities in addition to self-protection, such as self-optimization and self-healing. With enhanced swarm intelligence, IoT objects are capable of cooperating and sharing resources efficiently. This allows for solving numerous optimization problems, which are otherwise difficult to implement due to the large resources required. Self-optimization capabilities mean that SI can be used in many IoT applications, such as optimal node localization, optimal coverage control, and a wide variety of intelligent routings: shortest transportation path, best available channel at a point in time, minimum energy consumption.

The use of swarm intelligence supported by edge technologies (such as WSN), makes it possible to add more and more cognitive intelligence to the IoT objects, and at the same time add increasing swarm intelligence to the collaborative and connectivity space. Thus, IoT objects strive to improve to a higher level of local intelligence, close to human intelligence, in order to fulfil their function in a distributed manner, while the collective intelligence is centralized in order to solve problems that are more complex.

Swarm intelligence allows IoT to adopt a wide range of solutions already found in AI, data mining and robotics, so that IoT applications become more robust, flexible, adaptable, scalable and self-organized. The self-organization property allows for the formation of swarms of various shapes and sizes.

Each IoT object, which is part of the swarm has an agent with just enough knowledge about its object (such as position, speed) in order to engage the object in collaborative tasks with other objects in the swarm. Thus, IoT objects may be fixed or mobile and the IoT objects may enter and leave the swarm as necessary, without disturbing the meshing architecture of the IoT system. Self-healing systems are another application of IoT. The self-healing property is found in systems that detect and diagnose problems, and thus must embed some form of fault tolerance. Fault-tolerance based on SI implies the generation of alternative transportation paths and the recovery of faulty paths, so that the information is not lost and need not be retransmitted.

3.7.3 IoT Trust Framework

Common IoT threats are presented in [47] together with requirements to make the IoT secure, involving several technological areas. The common thread seems to be the need for end-to-end security.

Trust and usability are very important success factors for IoT, the security and privacy of which need to be addressed across all the IoT architectural layers and across domain applications. Performance, complexity and costs are all factors, which influence adoption in addition to those that engender trust. While important progress has been made and actions have been planned to address usability, there nevertheless remain a number of potential gaps in the overall "trust" framework.

The adoption of fine-grained authorization mechanisms allows for more flexible resource control and enables tolerance when fronting unknown risks. In addition, IP security protocol variants for the IoT with public-key-based cryptographic primitives in their protocol design such as Datagram TLS (DTLS), the HIP Diet EXchange (DEX), and minimal IKEv2, can fulfil the requirements of the IoT regarding scalability and interoperability. End-to-end authentication, integrity confidentiality and privacy are essential.

It is very important for all IoT objects to collaborate with each other and with the environment in order to generate the most appropriate clustering for the task at hand, whether that be optimizing functions, locating and isolating attacked objects, alleviating damage, or healing. Objects' trustworthiness is therefore an important feature, which must involve addressing issues such as security, user access, user credentials/authentication, privacy, disclosure, and transparency. Developing an IoT trust framework addressing security, privacy, and sustainability in IoT products and services, as well as emphasising, "security and privacy by design" as part of IoT product and application development and deployment, is an important research priority for IoT activities.

It is important to keep in mind that all the technologies must be tailored to the constraints of IoT scenarios and to the characteristics of IoT devices, including limited memory, computing resources, security and backup connectivity.

Block chain technology is useful as a transaction-processing tool that can address trust and security issues and move towards open source and security based on transparency allowing the democratization of trust. This is done by holding a record of every transaction made by every participant and having many participants verify each transaction, providing highly redundant verification and eliminating the need for centralized trust authorities.

3.8 IoT Enabling the Digital Transformation of Industry

IoT refers to an ecosystem in which applications and services are driven by data collected from devices that sense and interface with the physical world. Important IoT application domains span almost all major economic sectors: health, education, agriculture, transportation, manufacturing, electric grids, and many more. Proponents of IoT techniques see a world in which a bridge's structural weaknesses are detected before it collapses, in which intelligent transportation and resilient electrical grids offer pleasant and efficient cities for people to live and work in, and in which IoT-supported e-applications transform medicine, education, and business. The combination of network connectivity, widespread sensor placement, and sophisticated data analysis techniques now enables applications to aggregate and act on large amounts of data generated by IoT devices in homes, public spaces, industry and the natural world. This aggregated data can drive innovation, research, and marketing, as well as optimise the services that generated it. IoT techniques will effect large-scale change in how people live and work. A thing in IoT can be an inanimate object that has been digitised or fitted with digital technology, interconnected machines or even, in the case of health and fitness, people's bodies. Such data can then be used to analyse patterns, to anticipate changes and to alter an object or environment to realise the desired outcome, often autonomously. IoT allows for tailored solutions, both in terms of production and services, in all industry areas. IoT data analytics can enable targeted medical treatment or can determine what the lot-size for certain products should be, effectively enabling the adaptation of production processes as required. In the context of manufacturing this would enable greater use of customised outcomes rather than trying to predict mass market demand. The IoT can empower people in ways that would otherwise not be possible, for example by enabling independence for people with disabilities and specific needs, in an area such as transport, or helping meet the challenges associated with an ageing society. Those countries that anticipate the challenges while fostering greater use will be best placed to seize the benefits [6].

In order to address the totality of interrelated technologies the IoT technology ecosystem is essential and the enabling technologies will have different roles such as components, products/applications, and support and infrastructure in these ecosystems. The technologies will interact through these roles and impact the IoT technological deployment [35].

IoT ecosystems offer solutions comprising a large system beyond a platform and solve important technical challenges in the different verticals and across verticals. These IoT technology ecosystems are instrumental for the deployment of large pilots and can easily be connected to or build upon the core IoT solutions for different applications in order to expand the system of use and allow new and even unanticipated IoT end uses.

There is a need to adapt research and innovation policies across a broad range of sectors and applications with focus on exchanging the data from and among the things and IoT platforms in an interoperable format. This requires creating systems that cross vertical silos and harvest the data across domains, which unleashes useful IoT applications that are user centric, context aware, and are able to create new services and providing gains from improvements in the base components of IoT, such as optimised wireless communications, data processing, analytics, etc.

Swarm intelligence can inspire the design of new IoT security solutions. In order to render this technology for IoT, it has to be fitted according to the IoT needs and as such more work is needed to understand limitations as well as an effective and interactive way to promote the development of these designs.

In applying the research and innovation, recommendations is important to consider the good practices developed to help policy makers move ahead and promote the positive elements of the IoT while minimising challenges and ensuring broader goals, including the following [6]:

- Evaluate and assess the existing policies and practices to determine that are suitably supportive of the IoT, and do not constitute unintentional barriers to potential IoT benefits.

- Promote the use of global technical standards for the IoT developed by standards setting bodies or industry consortia in order to support the development of an interoperable IoT ecosystem, while stimulating the emergence of new systems, boosting innovation and reinforcing competitiveness.
- As the communication technologies evolve, evaluate spectrum resources to satisfy IoT needs, both current and future, as different elements of the IoT, from machines to edge devices, need a variety of spectrum resources that is fit for purpose.
- Promote skills to maximise opportunities in the labour market and support workers whose tasks become displaced by IoT-enabled and IoT Robotic Things and systems, with adjustment assistance and re-skilling programmes.
- Build trust in the IoT by managing digital security and privacy risks in line with the global and European regulations and practices and by developing a Trust IoT framework based on cross-border and cross-sector interoperability of policy frameworks in the context of DSM.
- Support and further develop open data frameworks that enable the reuse of government data sets and encourage industry to share their non-sensitive data for public benefit.
- Promote and support the development of identity for things to address numbering, discovery, identity and access management. Flexibility is needed for numbering as different services or IoT users may have different requirements.
- Encourage the exploitation of the project results, support the private sector innovation taking advantage of the IoT, and improve the conditions for the creation of start-ups and IoT business models that are built around the opportunities created by the IoT applications and large scale pilots.

Internet of Things Timelines

Table 3.5 Future technological developments

Development	2016–2020	Beyond 2020
Identification Technology	• Identity management • Open framework for the IoT • Soft Identities • Semantics • Privacy awareness	• "Thing/Object DNA" identifier • Context aware identification • Context aware anonymity

IoT Architecture Technology	• Network of networks architectures • IoT reference architecture developments • IoT reference architecture standardization • Adaptive, context based architectures • Self-X properties	• Cognitive architectures • Distributed context, location, and state-aware architectures
IoT Infrastructure	• Cross domain application deployment • Integrated IoT infrastructures • Multi-application infrastructures • Multi provider infrastructures	• Global, general purpose IoT infrastructures • Global discovery mechanism
IoT Applications	• Configurable IoT devices • IoT in farming/water production and tracing • IoT in manufacturing industry • IoT in industrial lifelong service and maintenance • IoT device with strong processing and analytics capabilities • Application capable of handling heterogeneous high capability data collection and processing infrastructures • IoT wearables • IoT in smart cities • IoT and arts	• IoT information open market • Autonomous and Connected Vehicles • Internet of Buildings • Internet of Energy • Internet of Vehicles • Internet of Lighting • Internet of Health • Internet of Robotic Things • Internet of Farming • Internet of Industrial Things • Cognitive Internet • Tactile Internet
Communication Technology	• Wide spectrum and spectrum aware protocols • Ultra-low power chip sets • On chip antennas • Millimetre wave single chips • Ultra-low power single chip radios • Ultra-low power system on chip	• Unified protocol over wide spectrum • Multi-functional reconfigurable chips • Ultra-low power, short range IoT networks

(*Continued*)

Table 3.5 Continued

Development	2016–2020	Beyond 2020
	• Low-power wide-area networks (LPWANs) • Narrowband IoT (NB-IoT)	
Network Technology	• Network context awareness • Self-aware and self-organizing networks • Sensor network location transparency • IPv6-enabled scalability	• Network cognition • Self-learning, self-repairing networks • Ubiquitous IPv6-based IoT deployment
Software and algorithms	• Goal oriented software • Distributed intelligence, problem solving • Things-to-Things collaboration environments • IoT complex data analysis • IoT intelligent data visualization • Hybrid IoT and industrial automation systems • IoT devices over-the-air (OTA) firmware and software updates	• User oriented software • The invisible IoT • Easy-to-deploy IoT SW • Things-to-Humans collaboration • IoT 4 All • User-centric IoT
Hardware	• Smart sensors (bio-chemical) • More sensors and actuators (tiny sensors) • Sensor integration with NFC • Home printable RFID tags	• Nano-technology and new materials
Data and Signal Processing Technology	• Context aware data processing and data responses • Energy, frequency spectrum aware data processing	• Cognitive processing and optimisation
Discovery and Search Engine Technologies	• Automatic route tagging and identification management centres • Semantic discovery of sensors and sensor data	• Cognitive search engines • Autonomous search engines

Power and Energy Storage Technologies	<ul><li>Energy harvesting (biological, chemical, induction)</li><li>Power generation in harsh environments</li><li>Energy recycling</li><li>Long range wireless power Wireless power</li><li>Zero Power Listen-Mode mechanisms</li></ul>	<ul><li>Biodegradable batteries</li><li>Nano-power processing unit</li></ul>
Security, Privacy and Trust Technologies	<ul><li>User centric context-aware privacy and privacy policies</li><li>Privacy aware data processing</li><li>Security and privacy profiles selection based on security and privacy needs</li><li>Privacy needs automatic evaluation</li><li>Context centric security</li><li>Homomorphic Encryption</li><li>Searchable Encryption</li><li>Protection mechanisms for IoT DoS/DdoS attacks</li></ul>	<ul><li>Self-adaptive security mechanisms and protocols</li><li>Self-managed secure IoT</li><li>Swarm intelligence</li><li>Artificial intelligence</li><li>Deep learning security mechanisms</li></ul>
Interoperability	<ul><li>Optimized and market proof interoperability approaches used</li><li>Interoperability under stress as market grows</li><li>Cost of interoperability reduced</li><li>Several successful certification programmes in place</li></ul>	<ul><li>Automated self-adaptable and agile interoperability</li><li>Plug'n'Play Interoperability</li></ul>
Standardisation	<ul><li>IoT standardization refinement</li><li>M2M standardization as part of IoT standardisation</li><li>Standards for cross interoperability with heterogeneous networks</li><li>IoT data and information sharing</li></ul>	<ul><li>Standards for autonomic communication protocols</li></ul>

Table 3.6   Internet of Things research needs

Research Needs	2016–2020	Beyond 2020
Identification Technology	• Convergence of IP and IDs and addressing scheme • Unique ID • Multiple IDs for specific cases • Extend the ID concept (more than ID number) • Electro Magnetic Identification – EMID	• Multi methods – one ID
IoT Architecture	• IoT layered architecture based on use cases from global scale applications, global interoperability, and interconnections of many trillions of things	• New algorithms, architectures, data structures and approaches to machine learning • Pervasive, secure IoT network architectures • Knowledge sharing IoT networks
IoT Infrastructure	• Application domain-independent abstractions and functionality • Cross-domain integration and management • Large-scale deployment of infrastructure • Context-aware adaptation of operation	• Self-management and configuration • Self-healing • Swarm intelligence and adaptation mechanisms
IoT Applications	• IoT information open market • Standardization of APIs • IoT device with strong processing and analytics capabilities • Ad-hoc deployable and configurable networks for industrial use • Mobile IoT applications for IoT industrial operation and service/maintenance • Fully integrated and interacting IoT applications for industrial use	• Building and deployment of public IoT infrastructure with open APIs and underlying business models • Mobile applications with bio-IoT-human interaction • Tactile Internet of Things • Internet of Robotic Things • Virtual reality things • Augmented Things Reality
IoT Platforms and Software Services for IoT	• IoT Platforms • Low-level device control and operations	• Fully autonomous IoT devices

	• IoT data acquisition, transformation and management • IoT application development • IoT Operating Systems • Quality of Information and IoT service reliability • Highly distributed IoT processes • Semi-automatic process analysis and distribution	• Integrated IoT cognitive platforms based on artificial intelligence including device monitoring, management, security, IoT data acquisition, event-driven logic, application programming, visualization, analytics
IoT Architecture Technology	• Code in tags to be executed in the tag or in trusted readers • Global applications • Adaptive coverage • Universal authentication of objects • Graceful recovery of tags following power loss • More memory • Less energy consumption • 3-D real time location/position embedded systems	• Intelligent and collaborative functions • Object intelligence • Context awareness • Cooperative position cyber-physical systems
Communication Technology	• Longer range (higher frequencies – tenths of GHz) • Protocols for interoperability • On chip networks and multi standard RF architectures • Multi-protocol chips • Gateway convergence • Hybrid network technologies convergence • 5G developments • Collision-resistant algorithms • Plug and play tags • Self-repairing tags	• Self-configuring, protocol seamless networks • Wide-area IoT networks
Network Technology	• Grid/Cloud network • Software defined networks • Service based network • Multi authentication • Integrated/universal authentication	• Need based network • Internet of Everything • Robust security based on a combination of ID metrics

(Continued)

Table 3.6   Continued

Research Needs	2016–2020	Beyond 2020
	• Brokering of data through market mechanisms • Scalability enablers • IPv6-based networks for smart cities	• Autonomous systems for nonstop information technology service • Global European IPv6-based Internet of Everything

Software and algorithms	• Self-management and control
	• Micro operating systems and IoT operating systems
	• Context aware business event generation
	• Interoperable ontologies of business events
	• Scalable autonomous software
	• Evolving software
	• Self-reusable software
	• Autonomous things:
	  o Self-configurable
	  o Self-healing
	  o Self-management
	• Platform for object intelligence
	• New application programming interfaces
		• Self-generating "molecular" software
		• Context aware software
		• Event stream processing
		• Distributed stream computing platforms (DSCPs)
		• Cognitive application programming interfaces
		• Data structures capable of learning and adapting to unique inbound data requirements over time

Hardware Devices	• Polymer based memory • IoT Processors • Ultra-low power EPROM/FRAM • Molecular sensors • Autonomous circuits • Transparent displays • Interacting tags • Collaborative tags • Zero Power Listen-Mode tags and sensors • Heterogeneous integration • Self-powering sensors • Low cost modular devices • Ultra-low power circuits • Electronic paper • Nano power processing units • Silent Tags • Biodegradable antennae	• Biodegradable circuits • Autonomous "bee" and "ant" type devices • Zero Power tags and sensors

Hardware Systems, Circuits and Architectures	• Multi-protocol front ends • Ultra-low cost chips with security • Collision free air to air protocol • Minimum energy protocols • Multi-band, multi-mode wireless sensor architectures implementations • Adaptive architectures • Reconfigurable wireless systems • Changing and adapting functionalities to the environments • Micro readers with multi standard protocols for reading sensor and actuator data • Distributed memory and processing • Low cost modular devices • Protocols correct by construction • IoT Device Management	• Heterogeneous architectures • "Fluid" systems, continuously changing and adapting
Data and Signal Processing Technology	• Common sensor ontologies (cross domain) • Distributed energy efficient data processing • Autonomous computing • Tera scale computing • Micro servers • Multi-functional gateways	• Cognitive computing Cognitive, software-defined gateways
Discovery and Search Engine Technologies	• Scalable Discovery services for connecting things with services while respecting security, privacy and confidentiality • "Search Engine" for Things • IoT Browser • Multiple identities per object • On demand service discovery/integration • Universal authentication	• Cognitive registries • Global IoT context aware and cognitive registry • Learning algorithms for search and discovery
Power and Energy Storage Technologies	• Paper based batteries • Wireless power everywhere, anytime • Photovoltaic cells everywhere • Energy harvesting • Power generation for harsh environments	• Biodegradable batteries

(Continued)

Table 3.6   Continued

Research Needs	2016–2020	Beyond 2020
Interoperability	• Dynamic and adaptable interoperability for technical and semantic areas • Open platform for IoT validation	• Self-adaptable and agile interoperability approaches
Security, Privacy and Trust Technologies	• Low cost, secure and high performance identification/authentication devices • Access control and accounting schemes for IoT • General attack detection and recovery/resilience for IoT • Cyber Security Situation Awareness for IoT • Context based security activation algorithms • Service triggered security • Context-aware devices • Object intelligence • Decentralised self-configuring methods for trust establishment • Novel methods to assess trust in people, devices and data • Location privacy preservation • Personal information protection from inference and observation • Trust Negotiation	• Cognitive security systems • Self-managed secure IoT • Decentralised approaches to privacy by information localisation • Swarm intelligence • Trusted IoT framework
Governance (legal aspects)	• Legal framework for transparency of IoT bodies and organizations • Privacy knowledge base and development privacy standards • Trusted IoT concept and principle • Governance by design	• Adoption of clear European norms/standards regarding Privacy and Security for IoT • Context aware governance
Economic	• Business cases and value chains for IoT • Emergence of IoT in different industrial sectors • Emergence of IoT ecosystems	• Integrated platforms • IoT ecosystems • Emergence of IoT across industrial sectors

Acknowledgments

The IoT European Research Cluster – European Research Cluster on the Internet of Things (IERC) maintains its Strategic Research and Innovation Agenda (SRIA), taking into account its experiences and the results from the on-going exchange among European and international experts.

The present document builds on the 2010, 2011, 2012, 2013, 2014 and 2015 Strategic Research and Innovation Agendas and presents the research fields and an updated roadmap on future R&D from 2016 to 2020 and beyond 2020.

The IoT European Research Cluster SRIA is part of a continuous IoT community dialogue supported by the EC DG Connect – Communications Networks, Content and Technology and international IoT stakeholders. The result is a lively document that is updated every year with expert feedback from on-going and future projects financed by the EC. Many colleagues have assisted over the last few years with their views on the IoT Strategic Research and Innovation agenda document. Their contributions are gratefully acknowledged.

List of Contributors

Abdur Rahim Biswas, IT, CREATE-NET, WAZIUP
Alessandro Bassi, FR, Bassi Consulting, IoT-A, INTER-IoT
Alexander Gluhak, UK, Digital Catapult, UNIFY-IoT
Amados Daffe, SN/KE/US, Coders4Africa, WAZIUP
Antonio Skarmeta, ES, University of Murcia, IoT6
Arkady Zaslavsky, AU, CSIRO, bIoTope
Arne Broering, DE, Siemens, BIG-IoT
Bruno Almeida, PT, UNPARALLEL Innovation, FIESTA-IoT, ARMOUR, WAZIUP
Carlos E. Palau, ES, Universitat Politècnica de Valencia, INTER-IoT
Charalampos Doukas, IT, CREATE-NET, AGILE
Christoph Grimm, DE, University of Kaiserslautern, VICINITY
Claudio Pastrone, IT, ISMB, ebbits, ALMANAC
Congduc Pham, FR, Université de Pau et des Pays de l'Adour, WAZIUP
Elias Tragos, GR, FORTH, RERUM
Eneko Olivares, ES, Universitat Politècnica de Valencia, INTER-IoT
Fabrice Clari, FR, inno TSD, UNIFY-IoT
Franck Le Gall, FR, Easy Global Market, WISE IoT, FIESTA-IoT, FESTIVAL

Frank Boesenberg, DE, Silicon Saxony Management, UNIFY-IoT
François Carrez, UK, University of Surrey, FIESTA-IoT
Friedbert Berens, LU, FB Consulting S.à r.l, BUTLER
Gabriel Marão, BR, Perception, Brazilian IoT Forum
Gert Guri, IT, HIT, UNIFY-IoT
Gianmarco Baldini, IT, EC, JRC
Giovanni Di Orio, PT, UNINOVA, ProaSense, MANTIS
Harald Sundmaeker, DE, ATB GmbH, SmartAgriFood, CuteLoop
Henri Barthel, BE, GS1 Global
Ivana Podnar, HR, University of Zagreb, symbIoTe
JaeSeung Song, KR, Sejong University, WISE IoT
Jan Höller, SE, EAB
Jelena Mitic DE, Siemens, BIG-IoT
Jens-Matthias Bohli, DE, NEC
John Soldatos, GR, Athens Information Technology, FIESTA-IoT
José Amazonas, BR, Universidade de São Paulo, Brazilian IoT Forum
Jose-Antonio, Jimenez Holgado, ES, TID
Jun Li, CN, China Academy of Information and Communications Technology,
    EU-China Expert Group
Kary Främling, FI, Aalto University, bIoTope
Klaus Moessner, UK, UNIS, IoT.est, iKaaS
Kostas Kalaboukas, GR, SingularLogic, EURIDICE
Latif Ladid, LU, UL, IPv6 Forum
Levent Gürgen, FR, CEA-Leti, FESTIVAL, ClouT
Luis Muñoz, ES, Universidad De Cantabria
Manfred Hauswirth, IE, DERI, OpenIoT, VITAL
Marco Carugi, IT, ITU-T, ZTE
Marilyn Arndt, FR, Orange
Markus Eisenhauer, DE, Fraunhofer-FIT, HYDRA, ebbits
Martin Bauer, DE, NEC, IoT-A
Martin Serrano, IE, DERI, OpenIoT, VITAL, FIESTA-IoT
Martino Maggio, IT, Engineering - Ingegneria Informatica Spa, FESTIVAL,
    ClouT
Maurizio Spirito, IT, Istituto Superiore Mario Boella, ebbits, ALMANAC,
    UNIFY-IoT
Maarten Botterman, NL, GNKS, SMART-ACTION
Ousmane Thiare, SN, Université Gaston Berger, WAZIUP
Payam Barnaghi, UK, UNIS, IoT.est

Philippe Cousin, FR, FR, Easy Global Market, WISE IoT, FIESTA-IoT, EU-China Expert Group

Philippe Moretto, FR, ENCADRE, UNIFY-IoT, ESPRESSO, Sat4m2m

Raffaele Giaffreda, IT, CNET, iCore

Roy Bahr, NO, SINTEF, UNIFY-IoT

Sébastien Ziegler, CH, Mandat International, IoT6

Sergio Gusmeroli, IT, Engineering, POLIMI, OSMOSE, BeInCPPS

Sergio Kofuji, BR, Universidade de São Paulo, Brazilian IoT Forum

Sergios Soursos, GR, Intracom SA Telecom Solutions, symbIoTe

Sophie Vallet Chevillard, FR, inno TSD, UNIFY-IoT

Srdjan Krco, RS, DunavNET, IoT-I, SOCIOTAL, TagItSmart

Steffen Lohmann, DE, Fraunhofer IAIS, Be-IoT

Sylvain Kubler, LU, University of Luxembourg, bIoTope

Takuro Yonezawa, JP, Keio University, ClouT

Toyokazu Akiyama, JP, Kyoto Sangyo University, FESTIVAL

Veronica Barchetti, IT, HIT, UNIFY-IoT

Veronica Gutierrez Polidura, ES, Universidad De Cantabria

Xiaohui Yu, CN, China Academy of Information and Communications Technology, EU-China Expert Group

Contributing Projects and Initiatives

IoT6, iCore, EURIDICE, IoT.est, OpenIoT, CuteLoop, BUTLER, IoT-A, SmartAgriFood, EAR-IT, ALMANAC, CITYPULSE, COSMOS, CLOUT, RERUM, SMARTIE, SMART-ACTION, SOCIOTAL, VITAL, WAZIUP, FESTIVAL, BeInCPPS, ESPRESSO, WISE IoT, FIESTA-IoT, iKaaS, ProaSense, MANTIS, ARMOUR, BIG IoT, VICINITY, INTER-IoT, symbIoTe, TAGITSMART, bIoTope, AGILE, Be- IoT, UNIFY-IoT.

List of Abbreviations and Acronyms

Acronym	Meaning
3GPP	3rd Generation Partnership Project
AAL	Ambient Assisted Living
AMR	Automatic Meter Reading Technology
API	Application Programming Interface
ARM	Architecture Reference Model
AWARENESS	EU FP7 coordination action Self-Awareness in Autonomic Systems

BACnet	Communications protocol for building automation and control networks
BAN	Body Area Network
BDI	Belief-Desire-Intention architecture or approach
Bluetooth	Proprietary short range open wireless technology standard
BUTLER	EU FP7 research project uBiquitous, secUre inTernet of things with Location and contExt-awaReness
CAGR	Compound annual growth rate
CE	Council of Europe
CEN	Comité Européen de Normalisation
CENELEC	Comité Européen de Normalisation Électrotechnique
CEP	Complex Event Processing
DNS	Domain Name System
DoS/DDOS	Denial of service attack Distributed denial of service attack
EC	European Commission
eCall	eCall – eSafety Support A European Commission funded project, coordinated by ERTICO-ITS Europe
EDA	Event Driven Architecture
EH	Energy harvesting
EMF	Electromagnetic Field
ERTICO-ITS	Multi-sector, public/private partnership for intelligent transport systems and services for Europe
ESOs	European Standards Organisations
ESP	Event Stream Processing
ETSI	European Telecommunications Standards Institute
EU	European Union
Exabytes	10^{18} bytes
FI	Future Internet
FI PPP	Future Internet Public Private Partnership programme
FIA	Future Internet Assembly
FIS 2008	Future Internet Symposium 2008

F-ONS	Federated Object Naming Service
FP7	Framework Programme 7
FTP	File Transfer Protocol
GS1	Global Standards Organization
Hadoop	Project developing open-source software for reliable, scalable, distributed computing
HC	Haptic Control
IAB	Internet Architecture Board
IBM	International Business Machines Corporation
ICANN	Internet Corporation for Assigned Name and Numbers
ICT	Information and Communication Technologies
iCore	EU research project Empowering IoT through cognitive technologies
IERC	European Research Cluster for the Internet of Things
IETF	Internet Engineering Task Force
INSPIRE	Infrastructure for Spatial Information in the European Community
IIoT	Industrial Internet of Things
IoB	Internet of Buildings
IoC	Internet of Cities
IoE	Internet of Energy
IoE	Internet of Everything
IoL	Internet of Lighting
IoM	Internet of Media
IoP	Internet of Persons, Internet of People
IoRT	Internet of Robotic Things
IoS	Internet of Services
IoT	Internet of Things
IoT6	EU FP7 research project Universal integration of the Internet of Things through an IPv6-based service oriented architecture enabling heterogeneous components interoperability
IoT-A	Internet of Things Architecture
IoT-est	EU ICT FP7 research project Internet of Things environment for service creation and testing
IoT-I	Internet of Things Initiative
IoV	Internet of Vehicles
IP	Internet Protocol

IPSO Alliance	Organization promoting the Internet Protocol (IP) for Smart Object communications
IPv6	Internet Protocol version 6
ITS	Intelligent Transportation System
KNX	Standardized, OSI-based network communications protocol for intelligent buildings
LOD	Linked Open Data Cloud
LTE	Long Term Evolution
M2M	Machine to Machine
MAC	Media Access Control data communication protocol sub-layer
makeSense	EU FP7 research project on Easy Programming of Integrated Wireless Sensors
MB	Megabyte
MIT	Massachusetts Institute of Technology
MPP	Massively parallel processing
NIEHS	National Institute of Environmental Health Sciences
NFC	Near Field Communication
NoSQL	not only SQL – a broad class of database management systems
OASIS	Organisation for the Advancement of Structured Information Standards
OEM	Original equipment manufacturer
OGC	Open Geospatial Consortium
OMG	Object Management Group
OpenIoT	EU FP7 research project Part of the Future Internet public private partnership Open source blueprint for large scale self-organizing cloud environments for IoT applications
Outsmart	EU project Provisioning of urban/regional smart services and business models enabled by the Future Internet
PAN	Personal Area Network
PET	Privacy Enhancing Technologies
Petabytes	10^{15} byte
PHY	Physical layer of the OSI model
PKI	Public key infrastructure

PPP	Public-private partnership
Probe-IT	EU ICT-FP7 research project Pursuing roadmaps and benchmarks for the Internet of Things
PSI	Public Sector Information
PV	Photo Voltaic
QoI	Quality of Information
RFID	Radio-frequency identification
SASO	IEEE international conferences on Self-Adaptive and Self-Organizing Systems
SDO	Standard Developing Organization
SEAMS	International Symposium on Software Engineering for Adaptive and Self-Managing Systems
SENSEI	EU FP7 research project Integrating the physical with the digital world of the network of the future
SIG	Special Interest Group
SLA	Service-level agreement/Software license agreement
SmartAgriFood	EU ICT FP7 research project Smart Food and Agribusiness: Future Internet for safe and healthy food from farm to fork
SmartSantander	EU ICT FP7 research project Future Internet research and experimentation
SOA	Service Oriented Approach
SON	Self-Organising Networks
SRIA	Strategic Research and Innovation Agenda
SI	Swarm Intelligence
SWE	Sensor Web Enablement
TC	Technical Committee
TI	Tactile Internet
USDL	Unified Service Description Language
UWB	Ultra-wideband
VR	Virtual Reality
W3C	World Wide Web Consortium
WSN	Wireless sensor network
Zettabytes	10^{21} byte
ZigBee	Low-cost, low-power wireless mesh network standard based on IEEE 802.15.4
Z-Wave	Wireless, RF-based communications technology protocol

Bibliography

[1] ISO/IEC JTC 1 – Information technology, Internet of Things (IoT) – Preliminary Report 2014, online at http://www.iso.org/iso/internet_of_things_report-jtc1.pdf

[2] Bluetooth, online at http://www.bluetooth.com

[3] ANT+, online at http://www.thisisant.com/

[4] It's confirmed: Wearables are the "next big thing", online at http://www.cnbc.com/2015/09/22/after-smartphones-wearable-tech-poised-to-be-next-big-thing.html

[5] Digital Economy Collaboration Group (ODEC), online at http://archive.oii.ox.ac.uk/odec/

[6] OECD Digital Economy Paper No 252, June 2016 – The Internet of Things – Seizing the Benefits and Addressing the Challenges, online at http://www.oecd-ilibrary.org/docserver/download/5jlwvzz8td0n.pdf?expires=1466330492&id=id&accname=guest&checksum=266AECBA35AAD3AC3F1BF7CAE3D8C409

[7] Wi-Fi Alliance, online at http://www.wi-fi.org/

[8] Z-Wave alliance, online at http://www.z-wavealliance.org

[9] Accenture. Are you ready to be an Insurer of Things?, online at https://www.accenture.com/_acnmedia/Accenture/Conversion-Assets/DotCom/Documents/Global/PDF/Strategy_7/Accenture-Strategy-Connected-Insurer-of-Things.pdf#zoom=50

[10] KPMG – security and the IoT ecosystem, online at https://www.kpmg.com/BE/en/IssuesAndInsights/ArticlesPublications/Documents/security-and-the-iot-ecosystem.pdf

[11] Is the Internet of Things Too Big to Protect? Not if IoT Applications Are Protected!, online at https://securityintelligence.com/is-the-internet-of-things-too-big-to-protect-not-if-iot-applications-are-protected/

[12] Gartner Says the Worlds of IT and Operational Technology Are Converging, online at http://www.gartner.com/newsroom/id/1590814

[13] Gartner IT Glossary: Operational Technology (OT) http://www.gartner.com/it-glossary/operational-technology-ot

[14] Gartner Inc. 2015. Gartner Hype Cycle, online at http://www.gartner.com/technology/research/methodologies/hype-cycle.jsp

[15] Gartner Inc. 2015. Newsroom. Gartner's 2015 Hype Cycle for Emerging Technologies, online at http://www.gartner.com/newsroom/id/3114217

[16] European Commission, ICT Standardisation Priorities for the Digital Single Market, Communication from the Commission to the European

Parliament, the Council, the European Economic and Social Committee and the Committee of the Regions, European Commission, 19-4-2016

[17] The Internet of Robotic Things, ABIresearch, AN-1818, online at https://www.abiresearch.com/market-research/product/1019712-the-internet-of-robotic-things/

[18] HART Communication Foundation, online at http://www.hartcomm.org

[19] IETF, online at https://www.ietf.org

[20] EnOcean Wireless Standard, online at http://www.enocean.com

[21] Gartner Identifies the Top 10 Internet of Things Technologies for 2017 and 2018, online at http://www.gartner.com/newsroom/id/3221818

[22] DASH7 Alliance, online at http://www.dash7.org

[23] RuBee, online at http://www.rubee.com/

[24] Botta, A., de Donato, W., Persico, V. and Pescapé, A., "Integration of Cloud computing and Internet of Things: A survey", Future Generation Computer Systems, Volume 56, March 2016, pp. 684–700.

[25] Mell, P. and Grance, T., "The NIST definition of Cloud computing", Natl. Inst. Stand. Technol., 53 (6), 2009, p. 50.

[26] Home Gateway Initiative (HGI), online at www.homegatewayinitiative. org

[27] Internet of Things, Services and People – IoTSP, ABB, online at http://new.abb.com/about/technology/iotsp

[28] Artemis IoE project, online at www.artemis-ioe.eu

[29] Wearables in healthcare, online at http://www.wearable-technologies. com/2015/04/wearables-in-healthcare/

[30] A Look at Smart Clothing for 2015, online at http://www.wearable-technologies.com/2015/03/a-look-at-smartclothing-for-2015/

[31] Best Smart Clothing – A Look at Smart Fabrics 2016, online at http://www.appcessories.co.uk/best-smart-clothing-a-look-at-smart-fabrics/

[32] Brunkhorst C., "Connected cars, autonomous driving, next generation manufacturing – Challenges for Trade Unions", Presentation at Industri-All auto meeting Toronto Oct. 14th 2015, online at http://www.industriall-union.org/worlds-auto-unions-meet-in-toronto

[33] The Internet of Things in Smart Buildings 2014 to 2020, Memoori report, online at http://www.memoori.com/portfolio/internet-things-smart-buildings-2014-2020/

[34] W. Arden, M. Brillouët, P. Cogez, M. Graef, et al., "More than Moore" White Paper, online at http://www.itrs.net/Links/2010ITRS/IRC-ITRS-MtM-v2%203.pdf

[35] O. Vermesan. The IoT: a concept, a paradigm, and an open global network. *Telit2market International*, Issue 10, February 2015, pp. 120–122, online at http://www.telit2market.com/wp-content/uploads/2015/02/telit2market_10_15_anniversary_edition.pdf

[36] Market research group Canalys, online at http://www.canalys.com/

[37] Platform INDUSTRIE 4.0 – Recommendations for implementing the strategic initiative INDUSTRIE 4.0, Final report of the Industrie 4.0 Working Group, online at, http://www.acatech.de/fileadmin/user_upload/Baumstruktur_nach_Website/Acatech/root/de/Material_fuer_Sonderseiten/Industrie_4.0/Final_report__Industrie_4.0_accessible.pdf, 2013

[38] P. C. Evans and M. Annunziata, Industrial Internet: Pushing the Boundaries of Minds and Machines, General Electric Co., online at http://files.gereports.com/wp-content/uploads/2012/11/ge-industrial-internet-vision-paper.pdf

[39] Cisco, "Securely Integrating the Cyber and Physical Worlds", online at http://www.cisco.com/web/solutions/trends/tech-radar/securing-the-iot.html

[40] H. Bauer, F. Grawert, and S. Schink, Semiconductors for wireless communications: Growth engine of the industry, online at www.mckinsey.com/

[41] ITU-T, Internet of Things Global Standards Initiative, http://www.itu.int/en/ITU-T/gsi/iot/Pages/default.aspx

[42] International Telecommunication Union – ITU-T Y.2060 - (06/2012) – Next Generation Networks – Frameworks and functional architecture models – Overview of the Internet of things

[43] IEEE-SA – Enabling Consumer Connectivity Through Consensus Building, online at http://standardsinsight.com/ieee_company_detail/consensus-building

[44] Mobile-Edge Computing – Introductory Technical White Paper, 2014, online at https://portal.etsi.org/Portals/0/TBpages/MEC/Docs/Mobile-edge_Computing_-_Introductory_Technical_White_Paper_V1%2018-09-14.pdf

[45] O. Vermesan, P. Friess, P. Guillemin, S. Gusmeroli, et al., "Internet of Things Strategic Research Agenda", Chapter 2 in Internet of Things – Global Technological and Societal Trends, River Publishers, 2011, ISBN 978-87-92329-67-7

[46] O. Vermesan, P. Friess, P. Guillemin, H. Sundmaeker, et al., "Internet of Things Strategic Research and Innovation Agenda", Chapter 2

in Internet of Things – Converging Technologies for Smart Environments and Integrated Ecosystems, River Publishers, 2013, ISBN 978-87-92982-73-5

[47] O. Vermesan, P. Friess, P. Guillemin, H. Sundmaeker, et al. Internet of Things Strategic Research and Innovation Agenda. O. Vermesan and P. Friess, Eds. *Internet of Things Applications – From Research and Innovation to Market Deployment.* Alborg, Denmark: The River Publishers, ISBN: 978-87-93102-94-1, 2014, pp. 7–142.

[48] SmartSantander, EU FP7 project, Future Internet Research and Experimentation, online at http://www.smartsantander.eu/

[49] Introducing Fujisawa SST – A town sustainably evolving through living ideas, Panasonic, online at http://panasonic.net/es/fujisawasst/

[50] H. Grindvoll, O. Vermesan, T. Crosbie, R. Bahr, et al., "A wireless sensor network for intelligent building energy management based on multi communication standards – a case study", ITcon Vol. 17, pg. 43–62, http://www.itcon.org/2012/3

[51] EU Research & Innovation, "Horizon 2020", The Framework Programme for Research and Innovation, online at http://ec.europa.eu/research/horizon2020/index_en.cfm

[52] Digital Agenda for Europe, European Commission, Digital Agenda 2010–2020 for Europe, online at http://ec.europa.eu/information_society/digital-agenda/index_en.htm

[53] Beecham Research Limited. Towards Smart Farming: Agriculture Embracing the IoT Vision, online at http://www.beechamresearch.com/download.aspx?id=40

[54] E. Guizzo. How Google's Self-Driving Car Works. IEEE Spectrum, online at http://spectrum.ieee.org/automaton/robotics/artificial-intelligence/how-google-self-driving-car-works

[55] Smartphone owners are ready for home and car IoT solutions, online at http://www2.deloitte.com/us/en/pages/technology-media-and-telecommunications/articles/internet-of-things-global-mobile-consumer-survey-infographic.html

[56] Freescale vision chip makes self-driving cars a bit more ordinary, online at http://www.cnet.com/news/freescale-vision-chip-makes-self-driving-cars-a-bit-more-ordinary/

[57] Mapping Smart City Standards – Based on a data flow model, online at http://www.bsigroup.com/LocalFiles/en-GB/smart-cities/resources/BSI-smart-cities-report-Mapping-Smart-City-Standards-UK-EN.pdf

[58] ISO/IEC JTC 1 – Information technology, Smart cities – Preliminary Report 2014, online at http://www.iso.org/iso/smart_cities_report-jtc1.pdf

[59] Report on Internet of Things Applications, AIOTI WG01, September 2015, online at https://ec.europa.eu/digital-single-market/en/news/aioti-recommendations-future-collaborative-work-context-internet-things-focus-area-horizon-2020

[60] Report on Innovation Ecosystems, AIOTI WG02, September 2015, online at https://ec.europa.eu/digital-single-market/en/news/aioti-recommendations-future-collaborative-work-context-internet-things-focus-area-horizon-2020

[61] Report on IoT LSP Standard Framework Concepts, AIOTI WG03, September 2015, online at https://ec.europa.eu/digital-single-market/en/news/aioti-recommendations-future-collaborative-work-context-internet-things-focus-area-horizon-2020

[62] Report on Policy Issues, AIOTI WG04, September 2015, online at https://ec.europa.eu/digital-single-market/en/news/aioti-recommendations-future-collaborative-work-context-internet-things-focus-area-horizon-2020

[63] Report on Smart Living Environment for Ageing Well, AIOTI WG05, September 2015, online at https://ec.europa.eu/digital-single-market/en/news/aioti-recommendations-future-collaborative-work-context-internet-things-focus-area-horizon-2020

[64] Report on Smart Farming and Food Safety Internet of Things Applications, AIOTI WG06, September 2015, online at https://ec.europa.eu/digital-single-market/en/news/aioti-recommendations-future-collaborative-work-context-internet-things-focus-area-horizon-2020

[65] Report on Wearables, AIOTI WG07, September 2015, online at https://ec.europa.eu/digital-single-market/en/news/aioti-recommendations-future-collaborative-work-context-internet-things-focus-area-horizon-2020

[66] Report on Analysis and Recommendations for Smart City Large Scale Pilots, AIOTI WG08, September 2015, online at https://ec.europa.eu/digital-single-market/en/news/aioti-recommendations-future-collaborative-work-context-internet-things-focus-area-horizon-2020

[67] Report on Smart Mobility, AIOTI WG09, September 2015, online at https://ec.europa.eu/digital-single-market/en/news/aioti-recommendations-future-collaborative-work-context-internet-things-focus-area-horizon-2020

[68] Report on Smart Manufacturing, AIOTI WG11, September 2015, online at https://ec.europa.eu/digital-single-market/en/news/aioti-recommenda tions-future-collaborative-work-context-internet-things-focus-area-hori zon-2020

[69] R. E. Hall, "The Vision of A Smart City" presented at the 2nd International Life Extension Technology Workshop Paris, France September 28, 2000, online at http://www.crisismanagement.com.cn/templates/blue/down_list /llzt_zhcs/The%20Vision%20of%20A%20Smart%20City.pdf

[70] Eurostat, Agriculture, forestry and fishery statistics, ISSN 1977-2262, 2013 edition, online at http://ec.europa.eu/eurostat/documents/3930297/ 5968754/KS-FK-13-001-EN.PDF/ef39caf7-60b9-4ab3-b9dc-3175b15 feaa6

[71] The Silver Economy as a Pathway for Growth Insights from the OECD-GCOA Expert Consultation, online at http://www.oecd.org/sti/the-silver-economy-as-a-pathway-to-growth.pdf

[72] Are You Implementing Internet of Things with the Right Database?, online at http://www.datastax.com/2014/06/implementinternet-of-things-with-the-right-database

[73] O. Vermesan, P. Friess, G. Woysch, P. Guillemin, S. Gusmeroli, et al., "Europe's IoT Stategic Research Agenda 2012", Chapter 2 in The Internet of Things 2012 New Horizons, Halifax, UK, 2012, ISBN 978-0-9553707-9-3

[74] Libelium, "50 Sensor Applications for a Smarter World", online at http://www.libelium.com/top_50_iot_sensor_applications_ranking#

[75] BUTLER, FP7 EU project, online at http://www.iot-butler.eu/

[76] Building smart communities, online at http://www.holyroodconnect.com/ tag/smart-cities/

[77] Using Big Data to Create Smart Cities, online at http://informationstrategy rsm.wordpress.com/2013/10/12/using-big-data-to-create-smart-cities/

[78] O. Vermesan, et al., "Internet of Energy – Connecting Energy Anywhere Anytime" in Advanced Microsystems for Automotive Applications 2011: Smart Systems for Electric, Safe and Networked Mobility, Springer, Berlin, 2011, ISBN 978-36-42213-80-9

[79] US$1.7bn raised in smart grid, battery and storage and efficiency sectors, online at http://www.metering.com/news/us1-7bn-raised-in-smart-grid-battery-and-storage-and-efficiency-sectors/

[80] M. M. Hassan, B. Song, and E. Huh, "A framework of sensor-cloud integration opportunities and challenges", in *Proceedings of the 3*[rd] *International Conference on Ubiquitous Information Management*

and Communication, ICUIMC 2009, Suwon, Korea, January 15–16, pp. 618–626, 2009.

[81] M. Yuriyama and T. Kushida, "Sensor-Cloud Infrastructure – Physical Sensor Management with Virtualized Sensors on Cloud Computing", NBiS 2010: 1–8.

[82] IBM and Samsung bet on Bitcoin Tech to save the Internet of Things, online at https://securityledger.com/2015/01/ibm-and-samsung-bet-on-bitcoin-to-save-iot/

[83] Mobile Edge Computing Will Be Critical For Internet-Of-Things And Distributed Computing, online at http://blogs.forrester.com/dan_bieler/16-06-07-mobile_edge_computing_will_be_critical_for_internet_of_things_and_distributed_computing

[84] Y. Bengio, Y. LeCun, "Scaling learning algorithms towards AI", *Large Scale Kernel Machines,* MIT Press, 2007

[85] Open Geospatial Consortium, Geospatial and location standards, online at http://www.opengeospatial.org.

[86] W3C Semantic Web, online at http://www.w3.org/

[87] European Commission, "Smart Grid Mandate, Standardization Mandate to European Standardisation Organisations (ESOs) to support European Smart Grid deployments", M/490 EN, Brussels, 2011

[88] Global Certification Forum, online at http://www.globalcertificationforum.org

[89] SENSEI, EU FP7 project, online at http://www.sensei-project.eu

[90] IoT-A, EU FP7 project, online at http://www.iot-a.eu

[91] IoT6, EU FP7 project, online at http://www.iot6.eu

[92] IoT@Work, EU FP7 project, online at https://www.iot-at-work.eu/

[93] Federated Object Naming Service, GS1, online at http://www.gs1.org/gsmp/community/working_groups/gsmp#FONS

[94] Ambient Assisted Living Roadmap, AALIANCE

4

Internet of Food and Farm 2020

**Harald Sundmaeker[1], Cor Verdouw[2], Sjaak Wolfert[2]
and Luis Pérez Freire[3]**

[1]ATB Institute for Applied Systems Technology Bremen, Germany
[2]Wageningen UR - LEI, The Netherlands
[3]GRADIANT, Spain

*"Surprise: Agriculture is doing more with IoT Innovation than most other
industries" Jahangir Mohammed [1]*

4.1 Global Food Production – Setting the Scene

Agriculture is of vital importance to feed Europe in a healthy way, while
Europe has also an important role in feeding the world. It is a large sector with
a big social and economic impact, e.g.:

- 43% of the EU's land area is being farmed [2];
- The food and drink industry is the largest manufacturing sector in the
  EU, representing 15% of EU manufacturing sector turnover [3];
- Agri-food logistics has 27% share in the EU road transport [4];
- Agri-food exports contribute to more than 7% to total EU exports in
  goods [5];
- Europe is the largest exporter of agri-food products in the world, EU28
  exports reached €122 billion in 2014 [5].

At the same time, the agri-food domain has to face very critical challenges, in
particular:

- Food security is a major issue, which will become even more urgent and
  critical in the next decades due to the expected increase of the world
  population and the growing economic wealth of emerging economies.
- In the meantime, we have already exceeded the carrying capacity of
  planet Earth with the current way of agricultural production. Further

globalization, climate change, a shift from a fuel-based towards a bio-based economy, and competing claims on land, fresh water and labour will complicate the challenge to feed the world without further polluting or overuse (Figure 4.1).

Figure 4.1 demonstrates the perpendicular paradigm shift needed. The right-bound dashed line indicates the current trend of an increasing food demand and the associated ecological footprint that is needed. Currently we are already at the point in which this footprint is already too large (two times our planet's carrying capacity). Our challenge will be: more than doubling of the agri-food production while at the same time at least halving our ecological footprint (Source: Wageningen UR).

- Increasing consumer concerns about food safety by the continuing sequence of food calamities, which have required massive product recalls, sometimes even on a European scale. Recent examples include the horsemeat scandal [6] and the E. coli outbreak [7].
- Agri-food supply chains are characterised by complex network structures where many small and medium enterprises (farms and parts of the processing industry) trade with huge multinationals in the input and retail sector. At this, agri-food products are often considered as commodities with cost-leadership as the dominant marketing strategy, resulting in low profit margins.

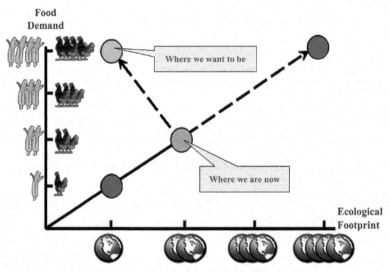

Figure 4.1   Food demand vs ecological footprint.

- Sustainable food chains are becoming 'license to deliver': roughly one-third of food produced for human consumption is lost or wasted globally [8] and food products account for an important part of the emissions produced by the transportation sector. The society does no longer accept the extremely high waste of food and the big CO2 footprint of food products.
- Growing attention for impact of food on health: consumers and society are increasingly aware that there is a strong relation between food consumption and so-called diseases of civilization, including obesity and food allergies.

Automation and mechatronics has enabled huge steps forward in in production efficiency, quality improvements and sustainability. For example, global crop yield increased by 77% between 1961–2007 [9] and the total greenhouse gas emissions of the primary production has been reduced by 23.8% in the period 1990–2012 [10]. The improvements are mainly accomplished by non-Internet technologies, such as mechanisation of field operations, breeding new varieties, and more environment-friendly cultivation techniques. Yet, the sector has to drastically increase productivity to feed the growing world population and to satisfy their changing food demands. This must be accomplished while at the same time agriculture is facing huge challenges in dealing with climate change, becoming more resource efficient, improving livestock conditions (animal welfare) and creating a circular economy, reducing waste, guarantying food safety and contributing to a healthy lifestyle of consumers.

The Internet of Things (IoT) is very promising for realizing new levels of control [11–13]. It is expected to be a powerful driver that will transform farming and food into smart webs of connected objects that are context-sensitive and can be identified, sensed and controlled remotely [14]. As such, we believe that IoT will be a real game changer in agriculture and the overall food chain that drastically improves productivity and sustainability, because it allows for [15]:

- Better sensing and monitoring of production, including farm resource use, crop development, animal behaviour and food processing;
- Better understanding of the specific farming conditions, such as weather and environmental conditions, animal welfare, emergence of pests, weeds and diseases, and creation of knowledge about appropriate management actions;
- More sophisticated and remote control of farm, processing and logistics operations by actuators and robots, e.g. precise application of pesticides

and fertilizers, robots for automatic weeding, or remote control of ambient conditions during transportation;
- Improving food quality monitoring and traceability by remotely controlling the location and conditions of shipments and products;
- Increasing consumer awareness of sustainability and health issues by personalised nutrition, wearables and domotics.

This situation is offering excellent opportunities for both the farm and food sector itself as well as for IoT providers.

4.2 Smart Farming and Food: Where We Are Right Now

The industrialisation of agriculture has expanded a lot in the previous decades. Farms and food companies are developing towards high-tech factories that are increasingly characterised by large-scale production and intensive use of technology. At the same time, those new IoT potentials are enabling new business models that were before impossible to realise. For example, the numerous startups that are following a basic trend to realise solutions that are offering fresh produce as well as to realise a very short supply chain from supplier to end-consumer [16], even leaving out steps of classical food chains. Collaboration of business partners is becoming more dynamic, competition for acquiring high quality produce at larger profit margin is increasing and delivery of information is rather a prerequisite for the realisation of innovative business models. Subsequently, data-rich management practices enabled by the IoT are of crucial importance for realising new levels of control resulting in a new jump in productivity and sustainability [17]. Some important advances in various domains (see Figure 4.2) pave the way for this breakthrough.

Precision Agriculture and Smart Farming

Precision Agriculture is about the very precise monitoring, control and treatment of animals, crops or m^2 of land in order to manage spatial and temporal variability of soil, crop and animal factors. In the previous decade so-called precision agriculture techniques have been introduced successfully [18]. For example, using satellite data, tractors can be very precisely located and steered making it possible to increase labour productivity especially for bigger machines: a 24 meter broad spraying machine requires advanced guidance equipment in order to avoid overlap and instability of spraying booms. By precise application of pesticides and fertilizers efficiency is increased, simultaneously reducing pollution. Another example from livestock

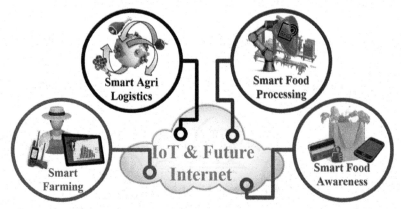

Figure 4.2 The various smart agri-food domains are increasingly integrated by the IoT and Future Internet technologies.

production is where RFID technology makes it possible for the amount of concentrate feed to be based on measured milk production for each individual cow. However, Precision Agriculture is adopted only by innovative farmers and the intelligent usage of precision farming data is still rather limited. At this, integration, ease of use and affordability are crucial bottlenecks [19]. Moreover, there are a lot of opportunities for the use of new sensors in the field and stables.

Smart Farming goes beyond the concept of Precision Agriculture by basing management tasks not only on location but also on data, enhanced by context- and situation awareness, triggered by real-time events. Real-time assisting reconfiguration features are required to carry out agile actions, especially in cases of suddenly changed operational conditions or other circumstances (e.g. weather or disease alert). These features typically include intelligent assistance in implementation, maintenance and use of the technology.

Figure 4.3 summarizes the concept of smart farming along the management cycle as a cyber-physical system. In this picture, it is already suggested that robots can play an important role in control, but it can also be expected that the role of humans in analysis and planning is further taken over by machines so that the cyber-physical cycle becomes fully autonomous. Of course, humans will still be involved in the whole process but probably at a much higher level of intelligence.

This role of farming robots is recognized in the forecasted supplies of service robots. It is expected that agriculture will be at the second place

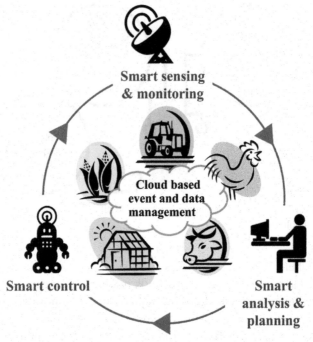

Figure 4.3 The cyber-physical management cycle of smart farming enhanced by cloud-based event and data management [21].

for demanding new service robots for professional use in the period of 2015–2018 [20]. This trend of an increased availability of high-tech devices will also facilitate to realise synergetic effects for IoT powered solutions, while different agri-food innovation fields can be considered as innovation drivers.

Smart Logistics: Food Traceability, Safety and Quality

Handling of food imposes the dimension of perishability upon all agri-food chain steps when planning and managing transport and storage. Logistic decisions need to be based on the underlying fact that quality characteristics are changing over time and due to environmental conditions. Unique identification of individual food items is rather complicated in terms of labelling, costs in relation to its value and real-world handling of food in cases, pallets and shipments. Therefore, produce and related packaging units are aggregated for being able to properly manage logistics, while virtualisation of shipments and its items, cases and pallets directly facilitates forwarding and storage.

The IoT provides sophisticated solutions for tracking and tracing as well as for remote management of shipments and products from production to the end-consumer. Figure 4.4 illustrates that such solutions allow supply chain actors to monitor, control, plan and optimize business processes remotely and in real-time through the Internet, based on virtual objects instead of observation on-site.

Food companies are obliged by law to trace products back to their origin and to track the ongoing location of products. This has forced companies worldwide to implement coordinated traceability systems along the food supply chain. However, food traceability is currently still often achieved by conventional systems, focusing on a single company or a specific part of the supply chain and using too basic technologies, e.g. product labelling, Electronic Data Interchange (EDI), email, paper trails [22]. Due to cost-benefit considerations, available RFID applications focus on container or pallet level, while single items are identified by barcodes. In most existing systems, traceability data are passed from one partner to the next one, while each partner records the supplier and customer of specific products ('one step forward

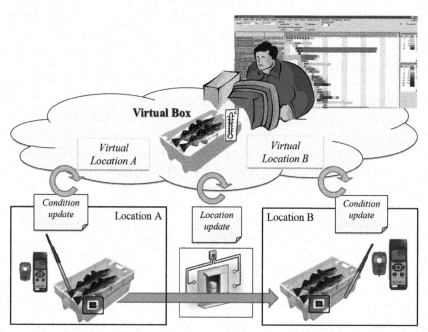

Figure 4.4 IoT enables the virtualization of agri-food supply chains: Example logistics of fresh fish, adapted from [14].

and one step back principle') [23]. There are some examples of Electronic Product Code Information System (EPCIS)-based traceability systems, which capture events of food items passing through a supply chain network, store these on one or more EPCIS-repositories and enable querying these events using appropriate security mechanisms [23–25]. Yet such solutions are still often implemented rather as closed systems than as open solutions serving dynamically changing business partners.

Sensor technologies are increasingly used for food safety and food quality management [26–28]. Temperature sensors for cold chain monitoring are common practice. Also sensors for other parameters including humidity, light, and ethylene are increasingly used. Furthermore, the majority of applied sensors are still fixed sensors and data loggers that are used to determine the causes of quality problems afterwards. The adoption of wireless sensors, especially Bluetooth, Zigbee, Wi-Fi and GPRS, is still in its infancy. The affordability of wireless sensors beyond temperature is a critical issue for wide-scale usage. Furthermore, many promising sensor technologies are still in an experimental stage of development, e.g. most chemical and bio sensors, electronic noses and Lab on a Chip [29, 30]. As a consequence, the microbiological quality is still measured in traditional laboratories, resulting in limited timeliness of food safety information. There are also solutions that add predictive analytics to quality monitoring solutions to determine remaining shelf life as well as to actively influence ripening processes, e.g. [28, 31–33].

Smart Food Processing and Manufacturing

Food processing plants are currently very much centrally controlled, which results in a limited flexibility. Application of IoT in food factories will be based on a more decentralised control concept. Machines become cyber-physical systems with embedded intelligence and local data processing and that communicate directly with other machines [34]. In such smart food factories, machinery is increasingly autonomous, managing its own service and maintenance requirements and adapting instantly to new production requirements. This approach is promoted by recent initiatives like the Industry 4.0 and Factory of the Future.

Smart Food Awareness

Consumers' trust in food, food production, the origin of food, and the actors involved is a core requirement for the functioning of European food markets and the competitiveness of industry involved. With the experience of scandals in mind, consumers increasingly expect transparency on which trust can be build. Transparency is not meant to know everything but to create awareness on the issues consumers are interested in, involving information on the safety and quality of products and processes, and increasingly on issues around environmental, social, and ethical aspects.

The IoT is rapidly changing the way of communication between consumers and food businesses. Having this in mind, the SmartAgriFood project has introduced a conceptual architecture for the Internet of Food and Farm [21], which is visualized in Figure 4.5. This can be considered as a kind of backbone for smart food awareness facilitating the feedforward as well as backward communication, finally required for making information available.

The majority of consumers currently have access to a wealth of food-related smartphone apps, including personalised nutrition advices, food traceability, recipes and purchasing support (including webshops). These applications are increasingly making use of connected sensors, wearables like smart watches, equipment at home (e.g. refrigerators, weighing machines) and outdoor equipment (e.g. in canteens, restaurants, super markets, fitness clubs). However, most consumer IoT applications related to food focus on specific functionalities and data, while the information exchange with other systems is limited.

4.3 Farming, Food and IoT: Where We Are Going

The previous section shows that, although there is a good technology base for application of IoT in farming and food, so far current IoT applications and technologies in the agri-food domain are still fragmentary and lack seamless integration. Especially more advanced solutions are in an experimental stage of development.

Operational applications are mainly used by a small group of innovators and still focus on basic functionalities at a high granularity level. However, we expect that this situation will change rapidly in the coming years. IoT technologies are currently maturing fast and most recently, IoT is in the spotlights of both users and technology providers in the farming and food domain. As a result, important advancements will be achieved, especially concerning:

- Ensuring the integration of existing IoT solutions with open IoT architectures, platforms and standards;
- Scaling-up the usage of interoperable IoT technologies beyond the innovators, including simplification of existing solutions to ensure attractiveness for the mainstream farmers and food companies;

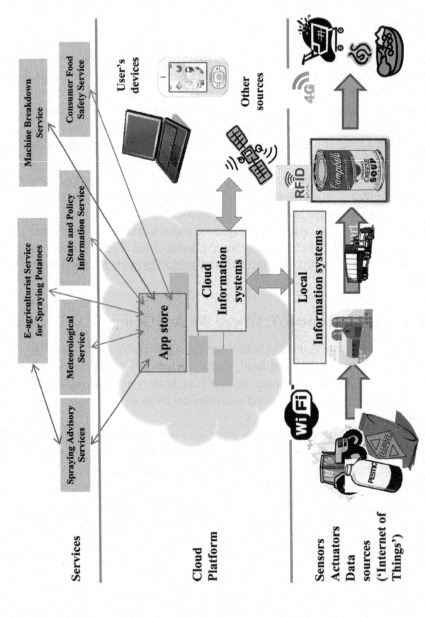

Figure 4.5   Conceptual architecture for the internet of food and farm as developed in the SmartAgriFood project.

- Further improvement of IoT technologies to ensure a broad usability in the diversity of the agri-food domain, e.g. different climate conditions, crop and soil types.

These technology developments are expected to drastically advance the development and application of the information technologies that were introduced in the previous section:

- *Precision Agriculture* will be extended to Smart Farming, in which a farm becomes a smart web of interoperable farm objects. A major improvement that will be added, is the seamless integration of sensing and monitoring, smart analyses and planning and smart control of farm operations for all relevant farm processes ('whole farm management perspective').
- *Tracking and Tracing* systems will develop towards end-to-end visibility and real-time tracking and tracing on a fine granularity level, e.g. up to individual products. Moreover, traceability will be increasingly integrated with smart sensing systems and consequently add data about product features, production methods, and ambient conditions.
- *Food Safety and Quality Management* systems will transform from a defensive, reactive approach towards a proactive approach, in which food chains can be monitored, controlled, planned and optimized remotely based on real-time information of a broad range of relevant parameters. To achieve this, more types of sensors will be put into practice, the timeliness of sensor information will be increased, the interoperability and thus end-to-end visibility of sensor data will be improved, more advanced remote control will be realized by implementing new actuators, and more intelligence will be added to food safety and quality management, for example: early warning in case of food incidents, rescheduling in case of unexpected food quality deviations and simulation of product quality based on ambient conditions (resulting in e.g. dynamic best-before dates).
- *Food Processing and Manufacturing* will increasingly be transformed into agile control systems in which processing machines function as autonomous connected objects with embedded intelligence. This will especially be achieved by integrating new and cost-effective sensors for real-time monitoring of processing activities, ensuring machine interoperability, and implementing algorithms for early detection of food safety and quality issues.
- *Consumer Food Awareness* will develop towards a fully consumer-centric approach that combines IoT technologies for different application

areas, including Smart Homes, Smart Shopping, and Smart Health and Leisure. These applications will combine food-related information from different stakeholders for personalised food intake advices.

IoT allows for the decoupling of physical flows and information aspects of farm and supply operations [14]. Farming processes and food supply chains can be monitored, controlled, planned and optimized remotely and in real-time based on virtual objects instead of observation on site.

Hence, farming and food will be transformed into smart webs of connected objects that are context-sensitive and can be identified, sensed and controlled remotely. This is expected to change agri-food processes in unprecedented ways, resulting in new control mechanisms and new business models such as:

- *Data-Driven Farming*: IoT will help farmers to change from 'management by gut feeling' towards 'management by facts', which is of crucial importance to survive its increasingly demanding business environment. The IoT sensing and connectivity technologies allow to feed decision-making tools with timely and accurate operational data.

- *Circular Economy/Green Farming and Food*: IoT will facilitate the control of using and distributing resources and it targets at a new dimension of symbiosis within the sector of food and farms. Collaboration with different industries can be facilitated that can supply their waste e.g. in form of heat, water, pressure or fertilizer. While also classical symbiotic systems like aquaponics will highly benefit from new IoT enabled control solutions facilitating distributed and autonomous operation.

- *Autonomous Farm Operations*: IoT will improve the connectedness and intelligence of farm automation. As such it will enable farm equipment to become autonomous, self-adaptive systems that can operate, decide and even learn without on-site or remote intervention by humans. Examples are automated precise control of farming equipment, weeding robots and self-driving tractors.

- *Demand-driven Farming*: IoT will enable farms to adjust and very accurately predict the volume and quality of supply by the precise and timely monitoring and control of production processes, also considering new interaction models that will communicate, feedback and predict the demand stemming from business as well as consumer side. As a consequence, farms can depart from the traditional supply-oriented, cost-price driven, anonymous approach to a value-based, information-rich approach in which demand and supply continuously are matched, both in offline and online distribution channels and combinations.

- *Outcome-based Agricultural Services*: IoT will significantly improve the possibilities to measure and control farm processes. As a consequence, farming will increasingly shift from competing through just selling products and services, to the ability to deliver quantifiable results that matter to customers, e.g. crop yield, energy saved or machine uptime [35].
- *Urban Farming*: IoT will support to situate fully-controlled indoor food production in urban areas close to the consumers. It will combine the advanced sensing and actuating technologies of IoT with new cultivation technologies for indoor farming (especially hydroponics, lighting).
- *Agile Food Factories*: IoT will enable a decentralised and flexible control concept for food processing by adding food sensors, local data processing and intelligence, and connectivity to food processing equipment.
- *Virtual Food Chains*: IoT will enable to virtualize food supply chain management, which allows for advanced remote (re)planning, monitoring and control capabilities and for new business models, for example: specialised virtual orchestrators that provide added value services and local-for-global trade by SMEs.
- *Personalised Nutrition*: IoT will allow for nutrition monitoring and personalised advices by using smartphones, that make use of connected sensors, wearables like smart watches, equipment at home (e.g. refrigerators, weighing machines) and outdoor equipment (e.g. in canteens, restaurants, super markets, fitness clubs).

4.4 Challenges

As seen in the previous sections, ICT technologies and IoT in particular are rapidly changing farming and the food industry. They have the potential to bring in the future, through large-scale deployments, huge benefits in the form of a more sustainable agriculture, ensuring food security with a lower environmental impact and guaranteeing healthier food production. However, reaping the full benefits will require overcoming certain IoT related challenges and barriers, both from technical and non-technical perspectives. At the same time, these difficulties bring new opportunities for technological development and value creation taking into account different types of stakeholders.

4.4.1 Technical Dimension

Speaking from a technical perspective, without trying to be exhaustive, the application of IoT to farming and food chains faces a series of challenges [15] such as:

Interoperability and Standardisation

Proprietary architectures, platforms and standards represent a barrier for the wide adoption of IoT in the agri-food sector due to the risks associated to vendor lock-in, incompatibility with other systems, etc. One of the challenges in the agri-food sector is to properly capture its particularities in the definition of new global, open standards and the alignment with existing standardisation initiatives from different stakeholders, stemming either from ICT (e.g. facilitated by the ETSI) or from agri-food (e.g. AEF, AgGateway, AgXML, GS1, ISO, UN/CEFACT) that need to be continuously aligned. In farming and food applications, one has to take into account farm management and traceability systems, agricultural machinery information exchange (including fleet management), and in general the specific data lifecycle (generation, collection, aggregation, visualization).

Enabling IoT Devices

Many of the benefits promised by IoT, including continuous and fine-grained monitoring of parameters and variables, will only come through technological breakthroughs such as the increase of computational power enabling edge computing/analytics, together with the drastic decrease of energy consumption in sensors and actuators to become (almost) energy-autonomous devices. The large-scale scope of farming applications also claims for more intelligence in the devices deployed in the field, including self-configuration and self-management capabilities. In traceability and food safety scenarios there is a clear challenge in developing new and cost-effective sensors and communication technologies, as for instance current biosensors, as well as RFID and NFC tags are not always viable (compared to the cost of the food product), in particular when targeting fine granularity, possibly at the individual product level. Further attention needs to be dedicated to the device characteristics, since food is rather a commodity with low profit margins and short lifetimes. Compared to tangible products from other sectors (e.g. clothing, furniture, multimedia), direct pairing of IoT with fresh produce is rather impossible often-requiring additional packaging. IoT potentials are not necessarily directly transferrable to food and farm, asking for additional efforts and costs, assuring that enabling IoT devices will neither be harmful to the environment, nor the consumers.

Enabling Network, Cloud and Communications

Connectivity is essential for making the best of IoT. However, IoT-intensive precision farming applications take place at food production (farms, aquaculture facilities), which are located in rural areas, where broadband coverage is still far too low, as only 4% of the rural European population has access to 4G connectivity, compared to 25% in towns and cities [36]. The massive deployment of smart devices will demand architectural changes in the communication networks (even at the Telco level) able to cope with specific data generation patterns and to rapidly adapt to changing traffic situations, thus bringing the need for advanced SDN/NFV technology. At the same time, agri-food is asking for IoT devices with a low power communication profile, even if this will reduce bandwidth and communication frequency, giving technologies like LoRa already quite some attention.

Information Services

Generation and collection of data is just the beginning in IoT applications. Extracting value from the data, in the form of meaningful and actionable information for the users, is the final goal. In this regard, although there are already good application examples, information services in the agri-food domain are still in an incipient stage. Short-term developments are mostly aimed at decision support systems, based usually on rules engines. More advanced data analytics, allowing for instance predictive modelling and production planning based on the demand (thus enabling *demand-driven farming*), are still a challenge in most agri-food applications. At this, object data has to be combined with a wealth of (3rd party) archives such as historical and forecasted meteorological data, satellite data, soil-, water- and air-analyses, logistic systems, and data on prices, logistics, retail, food service, and consumers, diets, etc. In this context, the usability of the information services is also of high interest: farm management systems should be easily adaptable to holdings of different sizes, and with a low learning curve for the user, while facilitating interoperability for horizontal and vertical collaboration of business partners in the agri-food chain.

Data Security

As explained in Section 4.3, farming and food chains (following the trend in other industries) are becoming more and more data-driven, so data becomes a precious asset. Indeed, the data captured by farming machinery potentially conveys a large amount of information, which is critical to the farmers, such as soil fertility and crop yield, so farmers must have strong guarantees on the protection of their data, in particular in cases where such data is stored (and possibly processed) in cloud-based services. As a consequence, many users are currently concerned about data ownership, privacy and security, which too often results in a lack of confidence and a 'wait-and-see' attitude. On the other hand, aggregation of data from different farms has the potential for generating huge added value. However, farmers must understand clearly the benefits they will get from such aggregation, as well as having the guarantee that their individual data is properly protected. In other words, Digital Rights Management solutions must be brought to the farming domain, in particular for scenarios of data aggregation and data sharing. This will also facilitate a promotion of open data initiatives for agri-food purposes as well as enabling an inter-sectorial collaboration.

From a technical security perspective, there are additional challenges to be considered in the domain of *trusted data*: the integrity and authenticity of the data generated and stored must be guaranteed. In traceability/safety applications, this is relevant to the origin of the product as well as the processing undergone, whereas in farming scenarios it is crucial, for example, in insurance-related issues. Trustworthiness requirements demand challenging solutions, such as low cost authentication mechanisms for devices/machines.

At the consumer side, security issues have to do more with personal data, thus bringing privacy at stake. For instance, IoT applications related to personalised nutrition imply privacy challenges because of personal and behavioural data captured from wearables, smartphones, etc.

IoT Platforms

As outlined in [37], there are numerous IoT platforms, stemming from open source initiatives as well as representing commercial IoT platforms. Besides the challenges with respect to governance, connectivity, fragmentation, inter-operability, and stakeholders, it is emphasised that the need for decision support at the application level to capitalise on the IoT, requires a loosely coupled, modular software environment based on APIs to enable endpoint data collection and interaction. This is specifically true for small- and medium-sized companies representing the majority in farming as well as parts in the food chain. A particular IoT empowered app might be enough to help solving a very particular problem. Apps could help to process or interpret data and make suggestions or give advice. For example: sensors in the field are

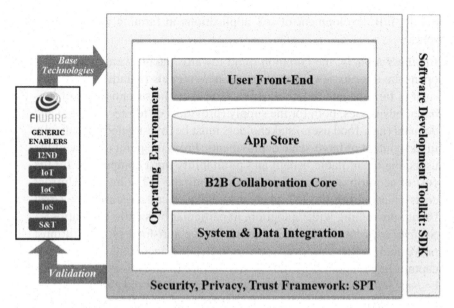

Figure 4.6 High-level picture of the FIspace architecture based on FIWARE GEs [38].

measuring the condition of the soil and consolidate this data in an app that is also predicting rain. As a consequence, the farmer is advised against spraying his field that day.

Therefore, the FIspace project has proposed an overarching architecture for enabling such kind of interactions, resulting in a multi-sided business-to-business collaboration platform [38], which is visualised in Figure 4.6. FIspace uses FIWARE Generic Enablers (GEs) but has two particular extensions for business collaboration: the App Store and the Real-Time B2B collaboration core. These key components are connected with several other modules to enable system integration (e.g. with IoT), to ensure Security, Privacy and Trust in business collaboration and an Operating Environment and Software Development Kit to support an ecosystem in which Apps for the FIspace store can be developed. The FIspace platform will be approachable through various type of front-ends (e.g. web or smartphone), but also direct M2M communication is possible.

4.4.2 Non-Technical Dimension

Besides the (non-exhaustive) technical challenges introduced above, from a non-technical perspective it is worth to mention other issues that are crucial towards the full development of IoT applications in farming and agri-food chains [15].

- *Business models*: the common trend towards data-driven value chains opens the door to new, disruptive business models in traditional sectors such as farming and food industries. However, the sustainability of IoT-based businesses, both for the supply (providers of IoT technology) and demand (agri-food users) stakeholders must be investigated, specifically in the context of large-scale deployments. From the point of view of the users, the quantifiable benefit and profitability must compensate for the cost of acquiring, operating and maintaining the IoT solutions. Upfront costs of acquiring the IoT platforms and services are currently a real barrier preventing wider adoption, in particular by small-sized farms.
- *Societal aspects*: IoT-based solutions for the agri-food sector still must prove their value massively to the users. IoT technologies enable to capture large amounts of data nearly in real time. However, data must be *beneficial to and useable for farmers* and all the stakeholders across the food chain. The benefits of the technology must be brought to real farming scenarios, thus dissemination and awareness are essential. An added difficulty in this regard is the heterogeneity of the agri-food value chain, including a large variety of holdings with many different sizes. In addition, to get the full benefit of IoT in farming applications it is essential that the users have certain digital skills. Currently, half of the EU population is not properly digitally skilled [36]. Thus, education and training in digital skills is essential to avoid creating a digital divide in the food and farm community. Farming cooperatives could play a key role in this regard.
- *Policy and regulations*: policies will play an essential role in the widespread deployment of IoT-based innovations in farming and food chains. In line with the Digital Single Market strategy [39] of the European Commission, they must help in lowering the existing barriers, which are slowing down the adoption of IoT. Just to mention a few which are directly related to some of the challenges/barriers mentioned above:
  - ○ Formulating clear security/privacy policies for protecting the farmers' data from unauthorised disclosure and for controlled and secure access to authorized third parties
  - ○ Supporting the faster rollout of broadband internet access in rural areas.
  - ○ Enhancing digital literacy skills and inclusion.

- **Stakeholder involvement**: We observe the changing roles of old and new software suppliers in relation to IoT, big data and agri-food. The stakeholder network exhibits a high degree of dynamics with new players taking over the roles played by other players and the incumbents assuming new roles in relation to agricultural data, information and knowledge. IoT in particular also entails organisational issues of farming and the supply chain. Further technological development may likely result in two supply chain scenarios from a stakeholder perspective. One with further integration of the linear supply chain in which farmers become franchisers. Another scenario in which farmers are empowered by IoT and open collaboration. The latter would enable also small stakeholders to easily switch between suppliers, share data with government and participate in short supply chains rather than integrated long supply chains.

4.5 Conclusions

The envisaged Internet of food and farm in the year 2020 is not just a rudimentary vision, but a path for research, technological development and most importantly for innovation. New IoT based solutions that are making an optimal usage of digital devices and the virtual world in challenging as well as harsh environments are promising a huge impact for agri-food business, technology providers and last but not least for all of us as consumers. Innovative solutions will pave the way helping to feed the global population, reducing emissions and resource usage per kg of food as well as avoiding empty trips of transport capacities with all its impact on CO_2 emissions and infrastructure maintenance. At the same time, consumers can become more aware of the overall agri-food chain that will help them to make informed decision when selecting specific produce. This can enable the opportunity to present consumers the full benefit of premium, organic and upcoming sustainable production methods as well as offer possibilities of better handling fair trade for farmers, hence facilitating their informed decisions.

As outlined before, the promising potentials of IoT technologies need to be based on an integrated usage of existing and mature methodologies and approaches that are already widely applied in the agri-food sector. Especially precision agriculture, traceability and food safety are cornerstones that are already part of the daily farm and food business practices. However, technology is still fragmented and data-rich management practices are not yet sufficiently in place, hampering to achieve a full extent of a symbiotic farm and food systems that are aiming at a continued increase of yields as achieved in the years before by non-Internet technologies. At the same time, IoT can be the key enabler to handle issues in relation to climate change, animal welfare and contributing to a healthy lifestyle of consumers.

Therefore, the potentials of an Internet of food and farm can enable e.g. autonomous farm operations, demand-driven farming and production as well as providing a personalised nutrition also based on virtual food chains. This will require technological and non-technological advances, while related difficulties will bring new opportunities for technological development and value creation. On the one hand, we need to work on data centric issues (e.g. interoperability, standardisation, security, service creation), while also finding new solutions to classical problems of using IoT devices and wireless communication in harsh and rural environments. On the other, heterogeneous types of stakeholders need to be empowered with the right digital skills, while policy and regulation have to be supportive in lowering barriers.

Finally, new and disruptive business models are in reach that will make use of the data-driven agri-food chain. However, the sustainability of IoT-based businesses, both for the supply (providers of IoT technology) and demand (agri-food users) stakeholders must be involved, specifically in the context of large-scale deployments for being able to mobilise a critical mass of end-users and validate the related benefits.

Bibliography

[1] Mohammed, J.; Surprise: Agriculture is doing more with IoT Innovation than most other industries. Founder and CEO of Jasper, December 7, 2014, Last accessed in May, 27th 2016; http://venturebeat.com/2014/12/07/surprise-agriculture-is-doing-more-with-iot-innovation-than-most-other-industries/

[2] Eurostat, March 2015, Land cover, land use and landscape. Last access on May, 27th 2016 http://ec.europa.eu/eurostat/statistics-explained/index.php/Land_cover,_land_use_and_landscape#cite_note-1.

[3] FoodDrinkEurope, Data & Trends European Food and Drink Industry 2014–2015.

[4] Eurostat, December 2015, Road freight transport by type of goods, http://ec.europa.eu/eurostat/statistics-explained/index.php/Road_freight_transport_by_type_of_goods.

[5] European Union, 2015, DG Agriculture & Rural Development: Agricultural Trade Policy Analysis unit, Agri-food trade in 2014: EU-US interaction strengthened, http://ec.europa.eu/agriculture/trade-analysis/map/2015-1_en.pdf.

[6] The Guardian; Horsemeat scandal: probe failure by authorities dates back to 1998. January, 21st 2014. Last accessed on May 27th 2016; http://www.theguardian.com/uk-news/2014/jan/21/horsemeat-scandal-p robe-failure-authorities-dates-back-1998-rotherham

[7] WHO Europe; Outbreaks of E. coli O104: H4 infection: update 30. July, 22nd 2011. Last accessed on May 27th 2016; http://www.euro.who. int/en/health-topics/disease-prevention/food-safety/news/news/2011/07/ outbreaks-of-e.-coli-o104h4-infection-update-30

[8] FAO, 2011, Global food losses and food waste – Extent, causes and prevention, Rome.

[9] FAO, 2012, cited in: Rabobank, 2015, Building a smarter food system: More productive, connected and sustainable.

[10] Eurostat, July 2015, Agriculture – greenhouse gas emission statistics, http://ec.europa.eu/eurostat/statistics-explained/index.php/Agriculture_-_greenhouse_gas_emission_statistics.

[11] Sundmaeker, H.; Guillemin, P.; Friess, P.; Woelfflé, S. (eds.), 2010, Vision and Challenges for Realising the Internet of Things, EC/CERP-IoT.

[12] Porter, M.E.; Heppelmann, J.E.; How Smart Connected Objects Are Transforming Competition. Harvard Business Review, November 2014.

[13] Sarni, W.; Mariani, J.; Kaji, J.; From Dirt to Data, The second green revolution and the Internet of Things. Deloitte Review, issue 18, 2016.

[14] Verdouw, C.N.; Wolfert, J.; Beulens, A.J.M.; Rialland, A., Virtualization of food supply chains with the internet of things, Journal of Food Engineering 176 (2015). – p. 128–136.

[15] AIOTI WG06 – Smart Farming and Food Safety Internet of Things Applications – Challenges for Large Scale Implementations, 2015.

[16] Sundmaeker, H.; Accelerating System Development for the Food Chain: a Portfolio of over 30 Projects, Aiming at Impact and Growth. 10th International European Forum on System Dynamics and Innovation in Food Networks (151st EAAE Seminar). Igls, Austria, February 2016.

[17] Poppe, K.J.; Wolfert, S.; Verdouw, C.; Verwaart, T.; Information and Communication Technology as a Driver for Change in Agri-food Chains. EuroChoices 12 (1) 2013, p. 60–65.

[18] EIP-AGRI Focus Group; Precision Farming; Final Report, November 2015.

[19] Aubert, B.A.; Schroeder, A.; Grimaudo, J.; IT as enabler of sustainable farming: an empirical analysis of farmers' adoption decision of precision agriculture technology. Decis. Supp. Syst., 54, 2012, pp. 510–520.

[20] Hägele, M.; Service robots are conquering the world. IFR Press Conference, Frankfurt, September 30th 2015.

[21] Wolfert, J., Sørensen, C.G. and Goense, D.; A future Internet collaboration platform for safe and healthy food from farm to fork, in Global Conference (SRII), 2014 Annual SRII. 2014, IEEE: San Jose, CA, USA. p. 266–273.

[22] Verdouw, C.N.; Sundmaeker, H.; Meyer, F.; Wolfert, J.; Verhoosel, J.; Smart Agri-Food Logistics: Requirements for the Future Internet, In: Dynamics in Logistics/Kreowski, H-J, Scholz-Reiter, B, Thoben, K-D, Berlin Heidelberg: Springer-Verlag, Lecture Notes in Logistics, 2013 – p. 247–257.

[23] Scholten, H.; Verdouw, C.N.; Beulens, A.J.M.; Vorst, J.G.A.J.v.d.; Defining and analysing traceability systems in food supply chains. In J. Bennett (Ed.), Advances in food traceability techniques and technologies. London, 2016: Elsevier.

[24] Verdouw, C.N.; Vucic, N.; Sundmaeker, H.; Beulens, A.J.M.; Future Internet as a Driver for Virtualization, Connectivity and Intelligence of Agri-Food Supply Chain Networks. International Journal on Food System Dynamics 4 (4), 2014 – p. 261–272.

[25] EuroPoolSystem; Delhaize in Belgium confirms its innovating role in fresh produce logistics. Last accessed in May 02nd 2016 http://www.europoolsystem.com/UploadBestanden/Article-Delhaize.pdf

[26] Heising, J.K.; Dekker, M.; Bartels, P.V.; Van Boekel, M.A.J.S.; Monitoring the Quality of Perishable Foods: Opportunities for Intelligent Packaging. Critical Reviews in Food Science and Nutrition 54, 5, 2013, p. 645–654.

[27] Jedermann, R.; Pötsch, T.; Lloyd, C.; Communication techniques and challenges for wireless food quality monitoring. Phil. Trans. R. Soc., 19, 2014.

[28] Verdouw, C.N.; Robbemond, R.M.; Verwaart, T.; Wolfert, J.; Beulens, A.J.M.; A reference architecture for IoT-based logistic information systems in agrifood supply chains. Enterprise information systems 2015. – 27 p, DOI:10.1080/17517575.2015.1072643.

[29] Loutfi, A.; Coradeschi, S.; Kumar Manib, G.; Shankarb, P.; Balaguru Rayappanb, J.B.; Electronic noses for food quality: A review, Journal of Food Engineering, 2015, 144, 103–111.

[30] Yoon, J-Y; Kim, B.; Lab-on-a-Chip Pathogen Sensors for Food Safety, Sensors 2012, 12(8), p. 10713–10741.

[31] Nollmann, S.; Alles Banane! Für eine bessere Qualität, Haltbarkeit und Ökobilanz von Frischeprodukten. VDI, VDE Mensch & Technik III/2011; p.23.

[32] Jedermann, R.; Dannies, A.; Moehrke, A.; Praeger, U.; Geyer, M.; Lang, W.: Supervision of transport and ripening of bananas by the Intelligent Container. In: 5th International Workshop Cold Chain Management, Bonn, Germany, University Bonn, 2013.

[33] FInish Accelerator; INVIVO – Shelf life prediction of perishable products. Last accessed on May 27th 2015; http://www.finish-project. eu/invivo/.

[34] Shinton, J.; Industry 4.0 in the food and beverage industry. EngineerLife, May, 27th 2015.

[35] World Economic Forum; Industrial Internet of Things: Unleashing the Potential of Connected Products and Services. January 2015; http:// www3.weforum.org/docs/WEFUSA_IndustrialInternet_Report2015.pdf

[36] European Commission, March 2015, Why we need a Digital Single Market? http://europa.eu/rapid/attachment/IP-15-4653/en/Digital_Single _Market_Factsheet_20150325.pdf

[37] Vermesan, O.; Friess, P.; Building the Hyperconnected Society – IoT Research and Innovation Value Chains, Ecosystems and Markets. IERC Cluster Book 2015.

[38] FIspace Project; Deliverable D200.2; FIspace Technical Architecture and Specification. October, 30th 2013, www.fispace.eu/publicdeliverables. html

[39] Communication from the Commission to the European Parliament, the Council, the European Economic and Social Committee and the Committee of the Regions: a Digital Single Market Strategy for Europe, 2015. http://g8fip1kplyr33r3krz5b97d1.wpengine.netdna-cdn.com/wp-content/uploads/2015/04/Digital-Single-Market-Strategy.pdf

5

Internet of Things Applications
in Future Manufacturing

John Soldatos[1], Sergio Gusmeroli[2], Pedro Malo[3] and Giovanni Di Orio[3]

[1]Athens Information Technology, Greece
[2]Politecnico di Milano, Italy
[3]UNINOVA & FCT NOVA, Portugal

5.1 Introduction

Future manufacturing is driven by a number of emerging requirements including:

- **The need for a shift from capacity to capability,** which aims at increasing manufacturing flexibility towards responding to variable market demand and achieving high-levels of customer fulfillment.
- **Support for new production models**, beyond mass production. Factories of the future prescribe a transition from conventional make-to-stock (MTS) to emerging make-to-order (MTO), configure-to-order (CTO) and engineer-to-order (ETO) production models. The support of these models can render manufacturers more demand driven. For example, such production models are a key prerequisite for supporting mass customization, as a means of increasing variety with only minimal increase in production costs.
- **A trend towards profitable proximity sourcing and production**, which enables the development of modular products based on common platforms and configurable options. This trend requires also the adoption of hybrid production and sourcing strategies towards producing modular platforms centrally, based on the participation of suppliers, distributors and retailers. As part of this trend, stakeholders are able to tailor final products locally in order to better serve local customer demand.

153

- **Improved workforce engagement**, through enabling people to remain at the heart of the future factory, while empowering them to take efficient decisions despite the ever-increasing operational complexity of future factories. Workforce engagement in the factories of the future is typically associated with higher levels of collaboration between workers within the same plant, but also across different plants.

The advent of future internet technologies, including cloud computing and the Internet of Things (IoT), provides essential support to fulfilling these requirements and enhancing the efficiency and performance of factory processes. Indeed, nowadays manufacturers are increasingly deploying Future Internet (FI) technologies (such as cloud computing, IoT and Cyber-Physical Systems (CPS) in the shop floor. These technologies are at the heart of the fourth industrial revolution (Industrie 4.0) and enable a deeper meshing of virtual and physical machines, which could drive the transformation and the optimisation of the manufacturing value chain, including all touch-points from suppliers to customers. Furthermore, they enable the inter-connection of products, people, processes and infrastructures, towards more automated, intelligent and streamlined manufacturing processes. Future internet technologies are also gradually deployed in the shopfloor, as a means of transforming conventional centralized automation models (e.g., SCADA (Supervisory Control and Data Acquisition), MES (Manufacturing Execution Systems), ERP (Enterprise Resource Planning)) on powerful central servers) towards more decentralized models that provide flexibility in the deployment of advanced manufacturing technology.

The application of future internet technologies in general and of the IoT in particular, in the scope of future manufacturing, can be classified in two broad categories:

- **IoT-based virtual manufacturing applications**, which exploit IoT and cloud technologies in order to connect stakeholders, products and plants in a virtual manufacturing chain. Virtual manufacturing applications enable connected supply chains, informed manufacturing plants comprising informed people, informed products, informed processes, and informed infrastructures, thus enabling the streamlining of manufacturing processes.
- **IoT-based factory automation**, focusing on the decentralization of the factory automation pyramid towards facilitating the integration of new systems, including production stations and new technologies such as sensors, Radio Frequency Identification (RFID) and 3D printing. Such

integration could greatly boost manufacturing quality and performance, while at the same time enabling increased responsiveness to external triggers and customer demands.

Within the above-mentioned categories of IoT deployments (i.e. IoT in the virtual manufacturing chains and IoT for factory automation), several IoT added-value applications can been supported. Prominent examples of such applications include connected supply chains that are responsive to customer demands, proactive maintenance of infrastructure based on preventive and condition-based monitoring, recycling, integration of bartering processes in virtual manufacturing chains, increased automation through interconnection of the shopfloor with the topfloor, as well as management and monitoring of critical assets. These applications can have tangible benefits on the competitiveness of manufacturers, through impacting production quality, time and cost. Nevertheless, deployments are still in their infancy for a number of reasons including:

- **Lack of track record and large scale pilots**: Despite the proclaimed benefits of IoT deployments in manufacturing, there are still only a limited number of deployments. Hence manufacturers seek for tangible showcases, while solutions providers are trying to build track record and reputation.
- **Manufacturers' reluctance**: Manufacturers are rather conservative when it comes to adopting digital technology. This reluctance is intensified given that several past deployments of digital technologies (e.g. Service Oriented Architectures (SOA), Intelligent Agents) have failed to demonstrate tangible improvements in quality, time and cost at the same time.
- **Absence of a smooth migration path**: Factories and production processes cannot change overnight. Manufacturers are therefore seeking for a smooth migration path from existing deployments to emerging future internet technologies based ones.
- **Technical and Technological challenges**: A range of technical challenges still exist, including the lack of standards, the fact that security and privacy solutions are in their infancy, as well as the poor use of data analytics technologies. Emerging deployments and pilots are expected to demonstrate tangible improvements in these technological areas as a prerequisite step for moving them into production deployment.

In order to confront the above-listed challenges, IoT experts and manufacturers are still undertaking intensive R&D and standardization activities. Such research is undertaken within the IERC cluster, given that several topics dealt within the cluster are applicable to future factories. Moreover, the Alliance for IoT Innovation (AIOTI) has established a working group (WG) (namely WG11), which is dedicated to smart manufacturing based on IoT technologies. Likewise, a significant number of projects of the FP7 and H2020 programme have been dealing with the application and deployment of advanced IoT technologies for factory automation and virtual manufacturing chains. The rest of this chapter presents several of these initiatives in the form of IoT technologies and related applications. In particular, the chapter illustrates IoT technologies that can support virtual manufacturing chains and decentralized factory automation, including related future internet technologies such as edge/cloud computing and BigData analytics. Furthermore, characteristic IoT applications are presented. The various technologies and applications include work undertaken in recent FP7 and H2020 projects, including FP7 FITMAN (www.fitman-fi.eu), FP7 ProaSense (http://www.proasense.eu), H2020 MANTIS (http://mantis-project.eu), H2020 BeInCPPS (http://www.beincpps.eu/), as well as the H2020 FAR-EDGE initiative. The chapter is structured as follows: The second section of the chapter following this introduction illustrates the role of IoT technologies in the scope of EU's digital industry agenda with particular emphasis on the use of IoT platforms (including FITMAN and FIWARE) for virtual manufacturing. The third section is devoted to decentralized factory automation based on IoT technologies. A set of representative applications, including applications deployed in FP7 and H2020 projects are presented in the fourth section. Finally, the fifth section is the concluding one, which provides also directions for further research and experimentation, including ideas for large-scale pilots.

5.2 EU Initiatives and IoT Platforms for Digital Manufacturing

5.2.1 Future Manufacturing Value Chains

The manufacturing Industry has recently evolved from rigid, static, hierarchical value chains to more flexible, open and peer-to-peer value ecosystems. Moreover, the added value produced by manufacturing (15% of the overall GDP (Gross Domestic Product) in the 28 EU Member States) has dramatically changed its pattern, where production and assembly of physical goods has constantly decreased its value added, in favor of pre-production and post-production activities.

The so-called SMILE challenge (Figure 5.1) is also emphasizing the role of ICT in this radical transformation of manufacturing value chains. In the central production stages, IoT is mostly at the service of Factory Automation and represents the major vehicle for connecting Real World (and its Cyber Physical Production Systems) with Digital-Virtual worlds in a green sustainable economy; in the pre-Production stages, closed loop collaboration ecosystems for new product-service design as well as Digital-to-Real world 3D printing ecosystems have been enabled by IoT and support creative economy; in the post-Production stages, new IoT-driven business models, supporting service- sharing- circular- economy, have been developed with success with the aim to compensate the loss of jobs derived from factory automation. For all the stages, it is necessary to proceed with the formation of new competencies and curricula centered on IoT and its related digital technologies, in order to attract young talents to Manufacturing and to up- re-skill existing workforce (blue and white collar workers).

Following paragraphs discuss the relevance of IoT in Manufacturing Value Chains, in consideration of two major events, which have characterized this year (i.e. 2016): the "Digitising EU" Industry policy communication and the enormous success of Industrie 4.0 initiatives and projects. A bi-directional convergence and innovation reference framework for digitizing EU Manufacturing value chains through IoT adoption is also proposed.

5.2.2 Recent EU Research Initiatives in Virtual Manufacturing

In commissioner Oettinger's speech at Hannover Messe on April 14th 2015, four main pillars for Europe's digital future were identified: i) Digital Innovation Hubs; ii) Leadership in platforms for Digital Industry; iii) Closing the digital skills gap and iv) Smart Regulation for Smart Industry.

On this basis, DG CNECT elaborated a yin-yang metaphor (Figure 5.2) to pictorially represent the two main challenges for achieving a strong EU Digital sector (against the GAFA US dominance) supporting a pervasive digitalization of EU industry. The "Collaborative Manufacturing and Logistics" FoF11 2016 call was partly focused on digital automation platforms for collaborative manufacturing processes, i.e. addressing together the first two pillars of Mr. Oettinger's speech: EU leadership in digital platforms to digitize EU manufacturing and logistics industries.

Many of the new FoF11 projects (currently under Grant Preparation phase in DG CNECT) are based on FIWARE and FITMAN Industrial IoT platform (Figure 5.3) and will bring new ideas and contributions to the IoT (IERC)

The "SMILE" challenge: European businesses must focus on high value added activities

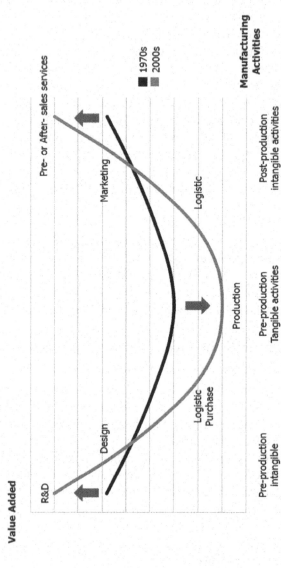

- Value creation in Manufacturing is progressively shifting **towards pre-production** (R&D and Design) and **post production** (marketing and Pre-or-After sales service) activities

Source: The European House - Ambrosetti re-elaboration on Bruegel data, 2014

Figure 5.1   Role of ICT in the transformation of manufacturing value chains.

Figure 5.2 Elements of industry digitization according to EU's vision.

Cluster in the next 2–3 years. They will also adhere and support AIOTI WG11 Smart Manufacturing, which is currently chaired by EFFRA (European Factories of the Future Research Association).

The convergence and coordination between IoT-focused projects (supervised by IoT European Platforms Initiative (EPI)) and other DG CNECT initiatives such as the aforementioned FIWARE (FITMAN), many FoF ICT projects such as I4MS BEinCPPS (based also on OpenIoT open source platform) and CPS/SAE initiatives represents the real challenge in the coming years for the IoT for Manufacturing domain of IERC.

In fact, the common research topic to be addressed by all the projects in the area of IoT-driven Digital Manufacturing Value Chains lies in the interrelation between the different aspects of IT (Information Technology, in this

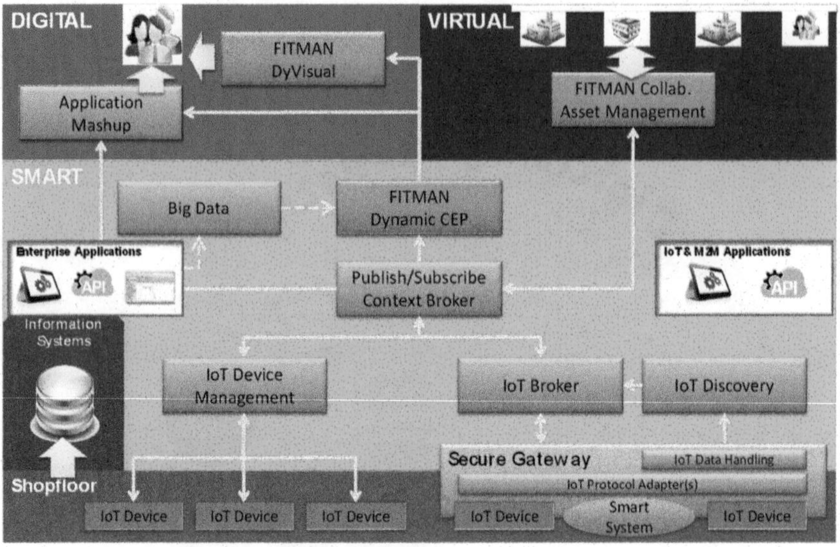

Figure 5.3   FP7 FITMAN and FIWARE projects include several IoT building blocks for digital manufacturing and virtual manufacturing chains.

case represented by IoT and CPS areas) and OT (Operation Technology) technology (in this case represented by Manufacturing Industries): stakeholders, reference architectures, platforms, physical and human resources, innovation and business models.

5.2.3 Levels of Manufacturing Digitisation

The recent EU communication about Digitising EU Industry of April 19th 2016 is exactly addressing this key topic, which is also the key topic for this interest group in IERC. The purpose of this Communication is to reinforce the EU's competitiveness in digital technologies and to ensure that every industry in Europe, in whichever sector, wherever situated, and no matter of what size can fully benefit from digital innovations. The communication aims at overcoming current barriers (e.g. high- vs. low-tech sectors, frontrunners vs. hesitators EU Countries, micro vs. small vs. large multinational enterprises), which prevents all EU manufacturing industries to achieve the following three progressive evolutionary levels of digitalization:

- **Digital Products**: driven by the development of the IoT to smart connected objects, it includes developments of markets like the connected car, wearables or smart home appliances.

- **Digital Processes**: driven by the development of IoT-enabled CPS, it includes Industrie 4.0, the further spread of automation in production and the full integration of simulation and data analytics over the full cycle from product design to end of life (circular economy).
- **Digital Business Models**: driven by service-oriented IoT-based business models, it includes the re-shuffling the value chains and blurring boundaries between products and services with the final aim to increase profitability by up to 5.3% and employment by up to 30% (in 2020).

According to the same EU communication, the achievement of this threefold objective is enabled by Digital Platforms, i.e. initiatives aiming at combining digital technologies, notably IoT, big data and cloud, autonomous systems and artificial intelligence, and 3D printing, into integration platforms addressing cross-sector challenges. In particular, leadership in IoT has recently seen an investment of the Commission in demand-driven large-scale pilots and lighthouse initiatives in areas such as smart cities, smart living environments, driverless cars, wearables, mobile health and agro-food. The investment will address notably open platforms cutting across sectors and accelerate innovation by companies and communities of developers, building on existing open service platforms, such as FIWARE. The accompanying staff-working document on IoT outlines among others standardisation and regulation challenges and opportunities for IoT and the role of the Alliance for IoT Innovations (AIOTI).

The Digitising Industry initiatives are aimed at a pervasive adoption of Information Technologies (IT) into Operations Technologies (OT), so they all implement the IT→OT way to do it. There is another perspective of the same topic (or the other side of the coin): the perspective of Manufacturing Industry, from OT→IT migration journey. This viewpoint is mostly represented by the German Industrie 4.0 and its subsequent EU-wide regional and national initiatives.

5.2.4 Industrie 4.0 Principles for CPS Manufacturing

The key focus of Industrie 4.0 is in the adoption of Cyber Physical Production Systems and in the consequent enablement of IoT and IoS applications (Figure 5.4).

Recently, several analysts identified so called "Industrie 4.0 readiness levels" to help manufacturing industries and especially SMEs to unleash the

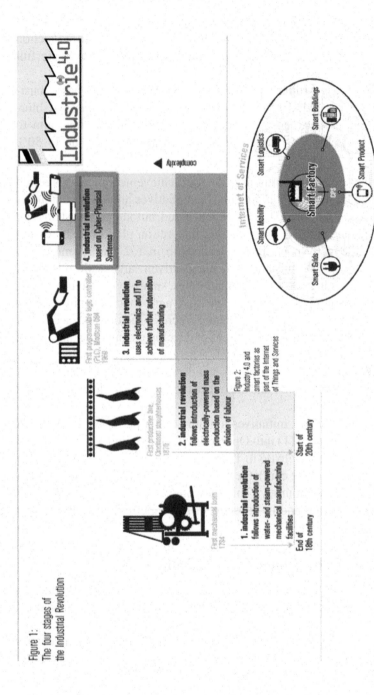

Figure 5.4   Industrial revolution steps: towards industrie 4.0.

full potential of digitalization of products, processes and business processes. In its most recent publications and in its speech at the World Manufacturing Forum 2016 in Barcelona, Max Blanchet, Senior Partner Automotive Industry, Process and Materials, Roland Berger, presented its model and the undoubtable benefits to Manufacturing Industry, deriving from a full adoption of seven key principles:

- From Mass Production to Mass Customisation.
- From volume Scale Effect to localized and flexible Units.
- From planned Make to Stock to dynamic Make to Order.
- From Product to Usage.
- From Cost driven to ROCE (Return on Capital Employed) driven.
- From Taylorism to flexible work organization.
- From hard working conditions to attractive work spaces.

The implementation of these seven principles in the manufacturing industry implies a migration of its resources towards IoT and the new IT.

As underlined before, the main research issue to be addressed in the Collaborative Digital Manufacturing Industry domain is the development of a bi-directional, win-win symbiotic model between IT and OT, in this case between IoT and Manufacturing. In this perspective, Europe is already playing a leading role worldwide in several so-called Key Enabling Technologies (KETs), such as micro- and nano-electronics, nanotechnology, industrial biotechnology, advanced materials, photonics, and advanced manufacturing technologies. When talking about bridging the Valley of Death between Research and Innovation, even at the small scale, such KETs are able per se to achieve a strong and immediate impact. In fact, in the first and second Phases of the I4MS initiative, some high-impact KETs (such as laser, robotics, High Performance Computing (HPC) simulation and CPS) have been and are being successfully transferred to Industry and SMEs in particular via a consistent ecosystem of local, small scale, almost independent champion experiments, grouped in Innovation Hubs. In the KETs domains, Technology Transfer approaches are based on increasing the TRL at the supply side, and on experimented lead-by-example success stories and best practices at the demand side, in order to give evidence to the whole ecosystem of the business benefits achieved. Once the effectiveness of the new KETs has been experimented on the field, the main barriers to their full adoption are mostly economic and financial: where and how to find the relevant resources to cover the sometimes huge investments required.

5.2.5 Digital Manufacturing and IoT Platforms

In terms of IoT, several Reference Architectures and Digital Platforms have been developed in diverse Research and Innovation actions at EU, National and Regional level, with the common aim to digitize manufacturing and logistics collaborative business processes. In the EC-funded FP7 and H2020 landscape, several R&I projects have been funded addressing the digitalization of manufacturing and logistics industries, not just in the Factories of the Future PPP (especially the recent C2Net, CREMA projects about Cloud Manufacturing and the Product Service System cluster), but also in other research environments such as Net Innovation (e.g. FIWARE for Industry, FITMAN and the Sensing Enterprise cluster), Cyber-Physical Systems (e.g. many H2020 ICT1 projects and the Smart Anything Everywhere cluster), IoT (e.g., 2015 Clusterbook of IERC Chapter 5, the AIOTI WG11 Smart Manufacturing and several "IoT for Manufacturing" workshops held at the recent and coming IoT WEEK events), Cloud Computing and Big Data (e.g. FIWARE PPP, IDS and BDVA) and even Technology Enhanced Learning (e.g. the TEL cluster for Manufacturing).

Many of these initiatives have been presented during several workshops organized along 2015 by DG CNECT. In particular, during the workshop of 5–6 October 2015, also initiatives not coming from EC-funded initiatives have been successfully demonstrated and discussed, such as Industry 4.0 RAMI (Reference Architecture Model Industrie 4.0) [1], Virtual Fort Knox, Industrial Data Space and the US Industrial Internet Reference Architecture IIRA. More recently, the newborn BEinCPPS Innovation Action in FoF I4MS phase II is aiming to integrate several of these platforms and to connect them via RAMI reference architecture also to National/Regional initiatives such as Virtual Fort Knox and Industrial Data Space.

However, the flourishing in EU of such an ecosystem of Research-driven IoT driven Platforms has not yet led to a successful and effective digitalization of all the aspects and resources of manufacturing and logistics industries involved in collaborative business processes: this is mainly due to the heterogeneity of the IT supply side (too many technologies and too many reference architectures, impossible to integrate into a common digital platform) and to the heterogeneity of the domains to be addressed and transformed in the Industry demand side (not just production systems, but also organizational, human resources, educational, business and just ultimately IT systems). Is IoT properly addressing the issues of data ownership and IPR management? Is Cloud Manufacturing a real opportunity for all manufacturing business processes, also those to be executed in real time? Have performance and security issues been solved? Is the Industrie 4.0 revolution based on CPPSs easy to be implemented in low-tech SMEs may be located in Eastern EU? If we look at the technological supply side, many of the above issues have been "solved" with advanced ICT solutions, but are the manufacturing industries ready for this revolution? Is there any Digital Platform to support their internal transformations, evolution to the new technologies?

In fact, when speaking of IoT-oriented Digital Platforms unleashing the full potential of collaborative business processes along the whole supply chain of manufacturing and logistics stakeholders, the process of digitizing industry requires complex, multi-domain and multi-disciplinary Large Scale Pilots (LSP) and cannot be effectively supported by simply putting in place mono-directional technology adoption initiatives based on increasing TRLs and Technology Transfer approaches.

In the case of Large Scale Pilots for Digital Platforms, TRLs are in fact not an absolute metric and often are dependent on contextual information, which cannot be ignored, such as size, sector, domain, digital literacy, location of the industries and their supply chains.

Moreover, as already said, often the activation of a huge ecosystem of Technology Transfer experiments is not the most effective option to create impact, in the presence of not well-prepared target industries and with respect to more holistic approaches like the creation of cross-domain interlinked regional ecosystems and Large Scale Pilots.

On the contrary, such a merely technology-driven approach risks to deepen the Digital Divide among industries, by favoring the excellence of leading edge champions, but offering inadequate support to lagging behind and low-tech industries. If not well prepared and conducted just via a mono-directional TRL-based technology transfer approach, Digital Automation risks to sharpen the divide between Eastern-Western EU Countries; between high- and low-tech sectors; between large multinational and local SMEs and mid-caps manufacturing industries.

More recently, in particular inside the AIOTI WG2 Innovation Ecosystem community led by PHILIPS and ELASTICENGINE, a new approach has been proposed: the appropriate way to measure the impact of these early adopter models would have to account for:

- The level of risk;
- The number of potential early adopters;
- Potential to yield data from early adoption; and finally
- The technology readiness.

We call them Market Adoption Readiness Levels (MARLs). This interesting approach for the very first time poses the increase of TRL as just one of the factors (the fourth one) to achieve innovation and not the unique way to impact. However, such an approach is mostly targeting consumer-centric and creative industries and needs substantial improvements and extensions to be applied to manufacturing domain, but in any case it is a quite promising starting point for a holistic approach to digital transformation of EU industry.

5.2.6 Maturity Model for IoT in Manufacturing

As indicated in the following picture, a Manufacturing Adoption Readiness Model:

- A first dimension considers the size and the investment capability of the manufacturing industry and its collaborative supply chain. Sometimes micro enterprises ecosystems are the fastest and most disruptive innovators, but they find difficult to create a real impact in the society, due to scarcity of investments. On the other side, large multi-national industries are seen as champions and archetypes for ICT-driven innovation, but often their migration processes are slow and bureaucratic. Economic feasibility and sustainability is the major maturity criterion addressed in this dimension;
- A second dimension considers the sector and industrial domain and its ICT awareness, where high-tech industries have already familiarity with certain technologies and young talented employees well prepared with respect to digital skills. On the contrary, low-tech industries heavily depend on knowledge and experiences of aging workers and engineers and the migration assumes in many cases the meaning of a generational knowledge transfer. Social sustainability is the major criterion addressed in this dimension.
- A third dimension considers the political and societal environment where the manufacturing supply chain operates. According to the Industry 4.0 readiness quadrant developed by Roland Berger consultants, four clusters of EU Countries could be identified according to two orthogonal variables: the Industry 4.0 readiness index (including degree of automation, workforce skills, innovation intensity and high value-added collaborative value networks) and the manufacturing vs. GDP ratio (the target 20% in 2020 for EU-28 countries according to former Commissioner Tajani's agenda). Hesitators are countries (such as Spain, Portugal and Estonia, plus presumably some EU associated countries like Serbia and Turkey)

with low readiness level and low GDP ratio; Traditionalist Countries (such as Italy, Poland, Croatia, Hungary, Slovenia) have a solid tradition in manufacturing – high GDP ratio – but a low readiness level and penetration of ICT into manufacturing industry; Potentialist Countries (such as UK, France, Denmark and the Netherlands) are good in ICT innovation but their manufacturing industry is not as developed as needed to achieve a deep societal impact; finally Frontrunners Countries (such as Germany, Ireland, Sweden and Austria) are leading edge environments where manufacturing digital innovation and societal impact are both well developed. Political sustainability is the major criterion addressed in this dimension.

A reference architecture for IoT-driven Digital Industry Collaborative Ecosystems could be inspired by the Industrie 4.0 RAMI, where hierarchical levels (from single components, to devices, to the whole connected world) are crossed with abstraction layers (from assets data, to information, to business knowledge) along the lifecycle of product typology and product instances (things lifecycle).

A first dimension of the IoT RAMI (Figure 5.5) (hierarchical technological levels, Y axis) considers technological assets and platforms, where Smart Networks, CPSs, IoT, Cloud, Big Data and Applications Marketplaces are considered.

This dimension is crossed with the second dimension (abstraction layers, Z axis) of the different types of Connected Factory resources involved in the migration processes: production resources, human resources, business resources, organizational resources and IT resources.

The third dimension (lifecycle, X axis) represents the evolution of digitalization patterns from smart products and production shop floors (digital inside, smart connected objects), to intelligent digitized M&L process (shop floor automation, energy optimization, preventive maintenance), to new business opportunities and innovation models (servitisation, sharing and circular economy), enabled by the migration to ICT.

In conclusion, the success of IoT-driven Digital Manufacturing Value Chains (Figure 5.6) depends on the simultaneous and coordinated implementation of a digitising Industry IT→OT roadmap aiming at increasing the TRL of IoT solutions and to extend the number of early adopters and success stories in manufacturing through Large Scale Pilots and of a migration to Industrie 4.0 OT→IT roadmap aiming at evolving manufacturing value chains' resources towards IoT and its technologies, by considering multi-dimensional maturity models and reference architectures derived from RAMI 4.0.

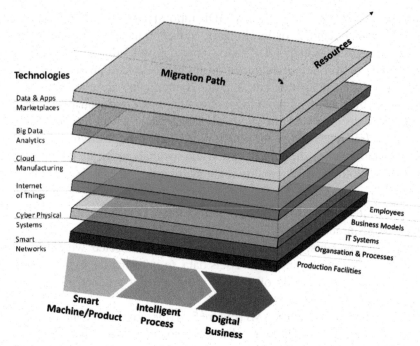

Figure 5.5   Dimensions of the reference architecture model industrie 4.0.

5.3 Digital Factory Automation

5.3.1 Business Drivers

Globalization has created a new and unprecedented landscape changing significantly the way manufacturing companies operate and compete: one of fierce
competition, shorter response time to market opportunities and competitor's
actions, increased product variations and rapid changes in product demand
are only some challenges faced by manufacturing companies of today. As in
other domains, production market has deeply felt the effects of globalization
on all different layers [2–4]. The increasing demand for new, high quality
and highly customized products at low cost and minimum time-to-market
delay is radically changing the way production systems are designed and
deployed. Success in such turbulent and unpredictable environment requires
production systems able to rapidly respond and adapt to changing markets
and costumer's needs. To capitalize on the key markets opportunities and
winning the competition for markets share, manufacturing companies are
caught between the growing needs for:

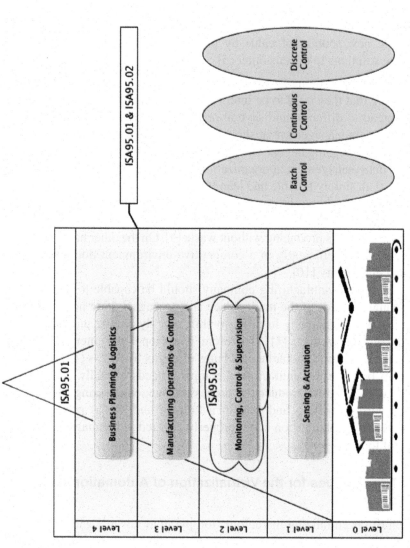

Figure 5.6   Manufacturing company functional hierarchical decomposition according to the ISA-95/IEC62264 standard.

- implementing more and more exclusive, efficient and sustainable production systems to assure a more efficient and effective management of the resources and to produce innovative and appellative customized products as quickly as possible with reduced costs while preserving product quality;
- creating new sources of value by providing new integrated product-service solutions to the customer [5].

In order to meet these demands, manufacturing companies are progressively understanding that they need to be internally and externally agile, i.e. agility must be spread to different and several areas of a manufacturing company from devices data management at shop floor level rising up to business data management while going beyond the individual company boundaries to intra enterprises data management at organization level. Therefore, agility implies being more than simply flexible and lean [6].

Flexibility refers to the ability exhibited by a company that is able to adjust itself to produce a predetermined range of solutions or products [7, 8], while lean essentially means producing without waste [9]. On the other hand, agility relates to operating efficiently in a competitive environment dominated by change and uncertainty [10].

Thus, an agile manufacturing company should be capable to detect the rapidly changing needs of the marketplace and propagate these needs to the lower levels of the company in order to shift quickly among products and models or between products [11]. Therefore, it is a top down enterprise wide effort that supports time-to-market attributes of competitiveness [12].

Thus, to be agile a manufacturing company needs a totally integrated approach i.e. to integrate product and process design, engineering and manufacturing with marketing and sale in a holistic and global perspective. Such holistic and global vision is not properly covered in the manufacturing company of today.

5.3.2 IoT Techniques for the Virtualization of Automation Pyramid

The vision of decentralizing the automation pyramid towards gaining additional flexibility in integrating new technologies and devices, while improving performance and quality is not new. Earlier efforts towards the decentralization of the factory automation systems have focused on the adaptation and deployment of SOA (Service Oriented Architecture) architectures for CPS and IoT devices [13]. However, SOA architectures tend to be heavyweight and rather inefficient for real-time problems, and therefore cannot be deployed in the shopfloor without appropriate enhancements. Furthermore, SOA deployments tend to focus on specific application functionalities and are not suitable for implementing shared situation awareness across all shopfloor applications. In recent years, the advent of edge computing architectures has provided a compelling value proposition for decentralizing factory automation systems, through the placement of data processing and control functions at the very edge of the network. Edge computing is one of the most prominent options for implementing IoT architectures that involve industrial automation and real-time control [14]. Nevertheless, the adoption of decentralized architectures (including edge computing) and IoT/CPS systems from manufacturers remains low for a number of reasons, including:

- **Lack of a well-defined and smooth migration path to distributing and virtualizing the automation pyramid**: The vast majority of manufacturers has heavily invested in their legacy automation architectures and are quite conservative in adopting new technologies, especially given the absence of a concrete and smooth migration path from conventional centralized systems to decentralized factory automation architectures. The virtualization of the automation pyramid could greatly benefit from a phased approach, which will facilitate migration, while also ensuring that the transition accelerates production, improves production quality and results in a positive ROI (Return-on-Investment).
- **IoT/CPS deployments and standards still in their infancy**: IoT/CPS deployments in manufacturing are still in their infancy. They tend to be overly focused on unidirectional data collection from sensors for remote monitoring purposes, while being divorced from the embedded and real-time nature of plant automation problems. At the same time, they tend to ignore the physical aspects of automation i.e. they pay limited emphasis on CPS aspects. Furthermore, despite the emergence of edge/fog computing architecture proposals for manufacturing (e.g., [14]) their implementation is still in its infancy.
- **Lack of shared situational awareness and semantic interoperability**: There is a lack of semantic interoperability across the heterogeneous components, devices and systems that comprise CPS-based automation environments for manufacturing. Distributed IoT/CPS components provide non-interoperable data and services, which is a set-back to creating sophisticated production automation workflows.
- **Lack of open, secure and standards-based platforms for decentralized factory automation**: The distribution of automation functions in the shopfloor is usually implemented on an ad-hoc fashion, which may

not comply with emerging architecture standards (such as the Reference Architecture Model Industrie 4.0). There is a lack of architectural blueprints for decentralized factory automation based on future internet technologies. Furthermore, emerging future internet platforms (such as FIWARE) have a horizontal nature and are not built exclusively for manufacturing domain (e.g., they do not address real-time requirements, complex security requirements and physical processes that characterize the FoF etc).

The advent of edge computing architectures, in conjunction with the emergence of IoT/CPS manufacturing as part of Industrie 4.0, promise to provide solutions for highly scalable distributed control problems which are subject to stringent real-time constraints. In particular, edge computing architectures are appropriate for processing or filtering large amounts of data at the edge of the network, as well as for performing large scale analysis of real-time data [15]. A digital automation platform based on edge computing and IoT technologies in the main objective of the H2020 FAR-EDGE project, which is currently in its contracting stage. This platform will comprise digital models of the plant, based on a proper compilation of reference/models and schemas for the digital representation of factory assets and processes (e.g., IEC-61987), notably reference schemas specified as part of RAMI. The platform will achieve distributed real-time control and semantic interoperability based on the replication and management of the state of the factory at the logical edges of the network and in a trustworthy way. The digital model of the factory and its secure sharing and distribution across the servers of the edge computing architecture will provide a foundation for the development of an operating system for factory automation, which could support a wide range of plant automation and control activities.

5.3.3 CPS-based Factory Simulation

The successful deployment of IoT analytics technologies in the shop floor hinges on the availability of digital datasets suitable for verification and validation of complex behaviors. The availability of such data cannot be taken for granted. The development of simulation services based on appropriate digital representations of plant could alleviate this limitation. Such simulation services need to consider the IoT architecture of the digital automation system along with the digital models of the representation of the plant.

The challenge lies not only to align the simulator with these models, but also to enable their sharing and synchronization across different automation processes.

5.3.4 IoT/CPS Production Workflows – Systems-of-Systems Automation

The next generation of industrial infrastructures are expected to be complex System-of-Systems (SoS) that will empower a new generation of industrial applications and associated services that are actually too hardly to implement and/or too costly to deploy [16]. There are several definition of a SoS in the literature, however, the definition that best fits the considered application context/domain is is the one provided in where SoS are defined as: [17] *"large-scale integrated systems that are heterogeneous and independently operable on their own, but are networked together for a common goal. The goal may be cost, performance, robustness, and so on"*. The state-of-the-art industrial automation solutions are known for their plethora of heterogeneous smart equipment encompassing distinct functions, form factors, network interfaces and I/O (Input/Output) specifications supported by dissimilar software and hardware platforms [18]. Such systems are designed, implemented and deployed to fulfill two main objectives:

- To convert raw materials, components, or parts into finished goods that meet a customer's expectations or specifications.
- To perform the conversion effectively and efficiently to guarantee a certain level of performance, robustness and reliability while keeping the costs low.

To do that, coordination, collaboration and, thus, integration and interoperability are extremely critical issues. Several efforts have been made towards the structural and architectural definition and characterization of a manufacturing company and its production management system as pointed in [19]. Among the others, the most popular and still practical applied is the set of definitions embodied into the ISA-95/IEC62264 standard (see Figure 5.6).

According to this standard manufacturing companies and their production systems (process plus factory) are organized into a five level hierarchical model also known as "automation pyramid". Besides this representation, the standard also provides a set of directives and guidelines for manufacturing operations management such as primary and secondary processes, quality assurance, etc. Even if the ISA-95/IEC62264 is the wider used approach for modelling manufacturing companies, nowadays it does not show all the intricacies of the applications, the communication protocols, and – more in general – of the several solutions present at each one of the five levels. Heterogeneity in terms of hardware and software – as well as – data distribution (transmission of information from several signal sources) and information processing are not fully covered by the ISA-95/IEC62264 standard. In fact it defines an information exchange framework to facilitate integration of business applications with the manufacturing control applications within a manufacturing company [20]. However, lower levels of the pyramid are not addressed implying that the automation pyramid – as it is – has significant limitations regarding the increased complexity of modern networked automation systems [21], and – in particular – it has limitations when it is used to support:

a) The integration of new technologies and devices and their lifecycle management;
b) The handling of the information flow along the overall automation pyramid from the lower level to the higher ones (company visibility);
c) The handling of the information flow coming from intelligent devices spread all over the living environment that could be used as fundamental feedback shared inside the automation pyramid.

The a), b), and c) limitations can be easily considered as different perspectives under the main umbrella of system integration research stream. In manufacturing, system integration can be addressed and instantiated at different levels of a company and, thus, with different levels of abstraction according to the context of application [22]. Each level presents a peculiar perspective about integration in general, and data integration in particular. Current technological trends in both industrial and living environments are pushing more and more to the idea of pervasive and ubiquitous computing while offering – at the same time – a huge opportunity to link information sources to information receivers/users. Future internet technologies – such as IoT and CPS – facilitate the deployment of advanced solutions in plant floor, as well as, day to day applications while promoting the meshing of virtual and physical devices and the interconnection of products, people, processes and infrastructures within the manufacturing value chain. The deployment of IoT/CPS-based systems is enabling the creation of a common virtualized space to facilitate the data acquisition process across multiple heterogeneous and geographically distributed data sources while facilitating the collaboration at large scale. It is necessary to comprehend that today's problem is no longer networking (protocols, connectivity, etc.) nor it is hardware (CPU/memory power is already there, at low-cost and low-power consumption) but rather it is on how to link disparate heterogeneous data sources – that are typically acquired from distinct vendors – to the specific needs and interaction forms of applications and platforms.

Designing and operating such complex systems requires from one side the presence of a generic reference model together with models, descriptions, guidance and specifications that can be used as key building blocks for deriving IoT/CPS-based architecture. From the other side, the increasing number of devices with advanced network capabilities is forcing the presence of intelligent middleware and more in general platforms where the whole enterprise is part of and where its internal components/devices can be easily discovered, added/removed/replaced and dynamically (re-)configured according to the business needs during the system operations and especially during the re-engineering interventions [16, 23, 24]. Several research initiatives and/or projects have been conducted to facilitate the interoperability of heterogeneous data sources. The IoT-A (http://www.iot-a.eu) project has addressed the IoT architecture and proposed a reference model as a response to a galaxy of of solutions somehow related to the world of intercommunicating and smart objects. These solutions show little or no interoperability capabilities as usually they are developed for specific challenges in mind, following specific requirements [25]. The Arrowhead (http://www.arrowhead.eu) project is aimed to provide an intelligent middleware/platform that can be used to allow the virtualization of physical machines into services. It includes principles on how to design SOA-based systems, guidelines for its documentation and a software framework capable of supporting its implementations. As a matter of fact, one of the main challenges of the Arrowhead project is the design and development of a framework to enable interoperability between systems that are natively based on different technologies. Most of the specifications are based on the models and outcomes provided by the FP7 IoT-A project.

5.4 IoT Applications for Manufacturing

5.4.1 Proactive Maintenance

As stated in [26], maintenance activities and procedures are always on high pressure from the top management levels of a manufacturing company to guarantee cost reduction while keeping the perfect working conditions of the machines and equipment installed in a production system, and in order to assure a certain degree of continuity in the productive process and – at the same time – the safety of the people that are part of it.

To do that, several policies and strategies for maintenance have been defined, developed and adopted, namely:

- Corrective Maintenance (CM);
- Preventive Maintenance (PM);

- Predictive Maintenance (PdM); and
- Proactive Maintenance (PrM).

In fact, maintenance owes its development essentially to the industrial progress in the recent centuries and to the growing need for manufacturing companies to be competitive [27].

Corrective Maintenance also called Run-to-failure reactive maintenance is the oldest policy and envisions the repair of a failure whenever it happens. It implies that a plant using run-to-failure management does not spend any money on maintenance until a machine or system fails to operate [28].

Preventive Maintenance is a time-driven policy and envisions the advanced definition of the time of intervention in order to anticipate the failure of complex system [27]. In preventive maintenance management, machine repairs or rebuilds are scheduled based on the mean time to failure (MTTF) statistic [28].

Predictive Maintenance also called condition-based maintenance is a policy that envisions the regular monitoring of machine and equipment conditions to understand their operating condition and schedule maintenance interventions only when they are really needed. In predictive maintenance management, machine repairs and/or rebuilds – i.e. maintenance interventions – are programmed in real-time avoiding unforeseen downtimes and their related implications [27]. As stated in [28]: Predictive maintenance is a philosophy or attitude that, simply stated, uses the actual operating condition of plant equipment and systems to optimize total plant operation. Finally, proactive maintenance is a totally policy that is not "failure" oriented like the others. As a matter of fact, proactive maintenance envisions not the minimization of the machine/equipment downtime but the continuous monitoring of the machine and equipment conditions with the main objective of identifying the root causes of a possible failure and/or machine breakdown and proactively schedule maintenance intervention to correct the abnormal values of the root causes. Thus, in proactive maintenance policy the minimization of the downtime is only the consequence of a strategy that is aimed to improve the machine/equipment health during its lifecycle and to assure overall high production system productivity, reliability, robustness while paradoxically reducing the number of maintenance intervention [29]. Proactive maintenance is a necessary state in the main path to effective maintenance. It has not been thought as an alternative to predictive maintenance but as a complementary approach to predictive maintenance in the direction of effective maintenance.

The successful implementation of proactive maintenance strategies strictly depends on the availability of an efficient and effective monitoring infrastructure that can gather relevant operational data from the machine/equipment combine and analyze this data to identify possible breakdowns and their root causes. However, current industrial monitoring and control solutions are extremely "bit-oriented" making hard and painful the process of predicting failures and detecting root causes. However, manufacturing companies are betting on the application of intelligent and more integrated monitoring and control solution to reduce maintenance problems, production line downtimes and reduction of production line operational costs while guarantying a more efficient management of the manufacturing resources [30].

In this context, IoT/CPS technologies can enable the design and development of advanced monitoring strategies and thus maintenance policies by adding additional monitoring capabilities to industrial machines and equipment providing in such a way the following functionalities:

- **Integration of secondary processes within the main control**: IoT/CPS based technologies can be deployed in order to provide more data about machine and equipment during their operation. Such information can be used to model the machine/equipment behavior for the sake of failures/breakdowns detection;
- **Modernization of low-tech production systems**: IoT/CPS based technologies can be deployed in low-tech production processes, i.e. production processes that are not natively ready for industry 4.0, and make them industry 4.0 compatible.
- **IT/OT Integration**: IoT/CPS technologies can easily provide data to all the layers of the automation pyramid enabling a true cross-layer integration.
- **Maintenance engagement**: IoT/CPS technologies can enable a better engagement of the maintenance department in the health of the overall production system.

5.4.2 Mass Customisation

The deployment of IoT technologies in virtual manufacturing chains and decentralized factory automation systems enables reduction of the production batch side and facilitates mass-customization. IoT devices can be deployed across the supply chain (e.g., even at retail locations) in order to obtain insights on customers' preferences. At the same time, the flexible integration of new technologies (such as stations, sensors and devices) facilitates the reduction of the batch size. Overall, IoT supports mass-customization across all points of the supply chain.

5.4.3 Reshoring

Decentralized IoT-based factory automation can enable European manufacturers to re-shore activities from low-labour countries back to the EU, which could have a positive impact on both employment and GDP (Gross Domestic Product). In particular, IoT enables reshoring through facilitating integration with advanced manufacturing technologies (e.g., IoT, 3D printing, robots, etc.) thus rendering manufacturing a far less labour intensive process. In this way, IoT enables a shift of manufacturing from low labour locations to locations with higher proximity to demand and innovation, which are the factors that will determine future locations for manufacturing.

5.4.4 Safe Human Workplaces and HMIs

The scope of Industrie 4.0 includes the dynamic adaptation, reconfiguration and streamlining of manufacturing processes. This reconfiguration occurs in response to variations in demand, while taking into account the status of the plant floor (e.g., machines, tools, control systems) in order to optimize the production workflow. Nevertheless, such adaptive and reconfigurable processes tend to neglect the human factor, given that they do not adequately take into account the employees' profile characteristics (such as age, disability, gender and skills). For example, RAMI, does not make any provisions for managing workforce profiles and human-centred processes. Likewise, Industrie 4.0 roadmaps are overly focused on technological issues and pay less attention on the ever important human and social factors (e.g., requirements for human-centred manufacturing). Overall, factory workers (include elderly and disabled workers) are still expected to fully adapt to the operations of machines and automation systems, even in cases of manufacturing workplaces with poor ergonomics.

In order to address these limitations, there is a need to devise technologies and processes that could invert the above loop i.e. put factory automation in the human workforce loop. Such technologies and processes could lead to a number of important benefits for manufacturers and for the society as whole, including: (A) Optimal integration among human and technical resources towards enhancing workforce performance and satisfaction; (B) Confronting the manufacturing skills gap; (C) Leveraging those individual worker capabilities that are most advantageous to the manufacturing process, while addressing important social factors (e.g., ageing and/or other handicapped groups) and ensuring health and safety at work; (D) Introduction of new flexible models of work and organization. Overall, there is a clear need for blending leading edge production automation technologies with state-of-the-art methodologies for human-centred processes and workplaces, including techniques for the adaptation of the physical workplace to the workers' characteristics and skills.

IoT technologies can enable manufacturers to support advanced ergonomics and novel models of work and organization through providing support for the following functionalities:

- **Human-centred production scheduling (notably in terms of workforce allocation)**: IoT technologies (such as RFID tags) can be deployed in order to provide access to the users' profile and context, thus enhancing conventional techniques for distributing tasks among workers in order to take into account the (evolving) profile and capabilities of the worker, including his/her knowledge, skills, age, disabilities and more.
- **Workplace Adaptation**: IoT devices such as sensors and PLCs (Programmable Logic Controllers) can provide the means for adapting the factory workplace operation and physical configuration (i.e. in terms of automation levels and physical world devices' configuration) to the characteristics, needs and capabilities of the workers, with a view to maximizing their performance and the overall productivity of the plant, while also maximising worker satisfaction.
- **Worker's engagement in the adaptation process**: IoT technologies can enable the comparison of the performance of a worker in a given task with the corresponding performance of skilled workers, in order to fine-tune the task distribution and workplace adaptation processes. Feedback on the performance of workers will be derived based on RFID tags and wearable devices, which can provide information about the workers stress, fatigue, sleepiness, and more.
- **Enhanced Workers' Safety and Well-Being**: The deployment and use of IoT wearables (such as Fitbit devices) can enable the tracking of the workers' activity levels. Fitbit data can be accordingly used to enhanced workers' safety and reduce healthcare and insurance costs for the manufacturers.

5.5 Future Outlook and Conclusions

5.5.1 Outlook and Directions for Future Research and Pilots

Previous paragraphs have presented a range of IoT technologies that can be used for streamlining manufacturing operations and for decentralizing factory automation. Despite the development of these technologies, there are still technological challenges especially in the following areas:

- **Security and Privacy**: IoT data in the shop floor varies in terms of volume and velocity, while including structured, unstructured and semi-structured data sources. At the same time, IoT deployments in manufacturing comprise multiple devices, which must be secured on the network. Holistic multi-layer approaches to security are therefore required in order to ensure safeguarding of personal data and control over the flow and exchange of sensitive information across the manufacturing chains and/or the shopfloor industrial network.
- **BigData Analytics**: Manufacturers need to convert data into actionable insight. Given the large volume of data, this is a significant challenge. The generation of business critical insights based on these data is still in its infancy, since data stemming from the manufacturing environment tends to be largely underutilized.
- **Adoption of Edge Computing**: Novel factory automation architectures have been largely based on the SOA and Intelligent Agents paradigm, in-line with standards such as the IEC 61499 Function Block. Emerging edge computing architectures have distinct advantages for the implementation of decentralized architectures, yet they have not been widely deployed yet.
- **Need for Standards-based Reference Implementations**: Recently, standards based organizations (such as the industrial Internet Consortium) have produced reference architectures for industrial automation and the integration of digital enterprise systems in the manufacturing chain. The provision of reference implementation of these standards will pave the way for their wider adoption and sustainable use by manufacturers.

Beyond the need to address these technical challenges, there is also a need for large-scale pilot deployments, which will combine several of the applications outlined in the previous section, in a way that considers their interactions and synergies. For example, proactive maintenance can give rise to effective production (re)scheduling, which could be also driven by information about customer demands (received via IoT devices). Likewise, IoT supported supply chain operations can drive the reconfiguration of production recipes, along with the scheduling of production. Moreover, the development of human centric workplaces requires the blending of adaptive human-centric processes (including appropriate HMIs (Human Machine Intefaces)) into IoT based factory automation architectures. Up to date, pilot deployments have been addressing only a fraction of the above listed applications, without a systematic consideration of their interactions under the prism of a standards-based reference implementation.

In addition to integrated pilots, large scale secure and privacy friendly deployments need to be evaluated in terms of quality, time and cost. Manufacturers need tangible evidence and benchmarks about IoT's ability to lead to improvements across these three axis, in order to provide a compelling proposition for adoption.

5.5.2 Conclusions

In this chapter, we have presented how IoT can transform manufacturing towards aligning to emerging trends such as proximity sourcing, support for flexible production models, human-centred manufacturing and more. We have also illustrated tangible deployments of IoT technologies based on recent FP7 and H2020 projects, notably projects focusing on the factories of the future. Despite these deployments, both technology and operational challenges exist. Reference implementation of standards compliant architectures for digital manufacturing based on IoT technologies could successfully address these challenges. Likewise, large-scale pilots combining the benefits of IoT deployments could also boost the confrontation of both technological and business/operational challenges.

Bibliography

[1] Vdi Vde Gesellschaft Mess und Automatisierungstechnik, Reference Architecture Model Industrie 4.0 (RAMI4.0), July 2015.
[2] D. F. Noble, Forces of Production: A Social History of Industrial Automation. Transaction Publishers, 2011.
[3] R. Narula, Globalization and Technology: Interdependence, Innovation Systems and Industrial Policy. Wiley, 2003.
[4] T. Levitt, "The globalization markets," The MITPress, vol. 249, 1993.
[5] S. Cavalieri and G. Pezzotta, "Product–Service Systems Engineering: State of the art and research challenges," Comput. Ind., vol. 63, no. 4, pp. 278–288, May 2012.

[6] G. Candido, "Service-oriented Architecture for Device Lifecycle Support in Industrial Automation," FCT-UNL, 2013.

[7] A. K. Sethi and S. P. Sethi, "Flexibility in manufacturing: A survey," Int. J. Flex. Manuf. Syst., vol. 2, no. 4, pp. 289–328, Jul. 1990.

[8] Y. P. Gupta and S. Goyal, "Flexibility of manufacturing systems: Concepts and measurements," Eur. J. Oper. Res., vol. 43, no. 2, pp. 119–135, Nov. 1989.

[9] S. S. Shah, E. W. Endsley, M. R. Lucas, and D. M. Tilbury, "Reconfigurable logic control using modular FSMs: Design, verification, implementation, and integrated error handling," in American Control Conference, 2002. Proceedings of the 2002, 2002, vol. 5, pp. 4153–4158 vol. 5.

[10] S. L. Goldman, R. N. Nagel, and K. Preiss, Agile competitors and virtual organizations: strategies for enriching the customer. Van Nostrand Reinhold, 1995.

[11] Y. Y. Yusuf, M. Sarhadi, and A. Gunasekaran, "Agile manufacturing: The drivers, concepts and attributes," Int. J. Prod. Econ., vol. 62, no. 1, pp. 33–43, 1999.

[12] P. Noaker, The Search for Agile Manufacturing. 1994.

[13] Spiess, P., Karnouskos, S., Guinard, D., Savio, D., Baecker, O., Souza, L. M. S. D., & Trifa, V. (2009, July). "SOA-based integration of the internet of things in enterprise services", IEEE International Conference on Web Services, ICWS 2009. (pp. 968–975).

[14] Industrial Internet Consortium. Industrial Internet Reference Architecture, June 2015.

[15] G. Orsini, D. Bade, W. H. Lamersdorf. Computing at the Mobile Edge: Designing Elastic Android Applications for Computation Offloading. 8th IFIP Wireless and Mobile Networking Conference (WMNC 2015), Munich, Germany, 2015.

[16] A. Colombo, T. Bangemann, S. Karnouskos, J. Delsing, P. Stluka, R. Harrison, F. Jammes, and J. L. Lastra, Eds., Industrial Cloud-Based Cyber-Physical Systems: The IMC-AESOP Approach, 2014 edition. New York: Springer, 2014.

[17] M. Jamshidi, System of Systems Engineering: Innovations for the Twenty-First Century. Wiley, 2009.

[18] G. Candido, C. Sousa, G. Di Orio, J. Barata, and A. W. Colombo, "Enhancing device exchange agility in Service-oriented industrial automation," in 2013 IEEE International Symposium on Industrial Electronics (ISIE), 2013, pp. 1–6.

[19] T. Bangemann, S. Karnouskos, R. Camp, O. Carlsson, M. Riedl, S. McLeod, R. Harrison, A. W. Colombo, and P. Stluka, "State of the Art in Industrial Automation," in Industrial Cloud-Based Cyber-Physical Systems, A. W. Colombo, T. Bangemann, S. Karnouskos, J. Delsing, P. Stluka, R. Harrison, F. Jammes, and J. L. Lastra, Eds. Springer International Publishing, 2014, pp. 23–47.

[20] M. Dassisti, H. Panetto, A. Tursi, and M. De Nicolo, "Ontology-based model for production-control systems interoperability," 2008.

[21] G. Pratl, D. Dietrich, G. P. Hancke, and W. T. Penzhorn, "A New Model for Autonomous, Networked Control Systems," IEEE Trans. Ind. Inform., vol. 3, no. 1, pp. 21–32, Feb. 2007.

[22] L. M. Camarinha-Matos and H. Afsarmanesh, "Brief Historical Perspective for Virtual Organizations," in Virtual Organizations, L. M. Camarinha-Matos, H. Afsarmanesh, and M. Ollus, Eds. Springer US, 2005, pp. 3–10.

[23] A. W. Colombo and S. Karnouskos, "Towards the factory of the future: A service-oriented cross-layer infrastructure," ICT Shap. World Sci. View Eur. Telecommun. Stand. Inst. ETSI John Wiley Sons, vol. 65, p. 81, 2009.

[24] J. Barata, Coalition Based Approach For ShopFloor Agility. Amadora – Lisboa: Orion, 2005.

[25] IoT-A, "Initial Architectural Reference Model for IoT," D1.2, Jun. 2011.

[26] G. Palem, "Condition-based Maintenance Using Sensor Arrays and Telematics," Int. J. Mob. Netw. Commun. Telemat. IJMNCT, vol. 3, no. 3, Jun. 2013.

[27] L. Fedele, Methodologies and Techniques for Advanced Maintenance. Springer-Verlag London Limited, 2011.

[28] K. Mobley, An Introduction to Predictive Maintenance. Elsevier Science Ltd, 2002.

[29] J. Levitt, Complete Guide to Preventive and Predictive Maintenance. Industrial Press Inc., 2011.

[30] G. Di Orio, "Adapter module for self-learning production systems," FCT-UNL, 2013.

6

Trusted IoT in the Complex Landscape of Governance, Security, Privacy, Availability and Safety

Elias Z. Tragos[1], Jorge Bernal Bernabe[2], Ralf C. Staudemeyer[3],
Jose Luis Hernandez Ramos[2], Alexandros Fragkiadakis[1],
Antonio Skarmeta[2], Michele Nati[4] and Alex Gluhak[4]

[1]FORTH-ICS, Greece
[2]Universidad de Murcia, Spain
[3]University of Passau, Germany
[4]Digital Catapult, United Kingdom

"Trust but verify". Ronald Reagan

6.1 Introduction

The Internet of Things (IoT) has attracted a lot of attention the last decade due to the unprecedented opportunities it provides for economic growth and for improving the quality of life of citizens. The advances in the IoT domain have been quite important and especially in the areas of IoT hardware, data and context extraction, service provisioning and service composition, cognition, interoperability and extensibility. Considering these advances, the IoT technologies are being considered quite mature for being deployed in real world environments and this has already been done in many cities around the world. Thousands of smart devices are now operating in cities, gathering information and providing smart applications for e.g. environmental monitoring, energy management, traffic management, smart buildings and smart parking [1, 2]. These devices are equipped with intelligence and are able to monitor and control physical objects, thus creating a new "Cyber-Physical" world [3].

The latest advances in the manufacturing engineering has allowed the minimization of the size of IoT devices so that they are not easily noticed.

Additionally, the humans are nowadays so familiar with computers and small devices that do not even pay attention to them, considering them as a part of their everyday lives. These two facts are proving how true for IoT was the projection from Marc Weiser back in 1991 when he described the "computer of the 21st century" using the phrase [4]:

> The most profound technologies are those that disappear. They weave themselves into the fabric of everyday life until they are indistinguishable from it.

It is easily understood that this phrase can characterize the IoT technology and its future inclusion within the everyday activities of the humans. The projection is that people will become so familiar with IoT that they will consider the technology as part of their lives. Although this shows the huge potential of IoT and its power, it raises significant concerns regarding security, privacy and safety. Imagine thousands and millions of small, unnoticeable devices spread around in city areas and within buildings monitoring and logging the everyday activities of people and controlling physical objects (doors, windows, cars, traffic lights, etc.) [5]. This can be quite worrying for the privacy of the people if the IoT systems are not designed to be secure and privacy preserving. However, IoT is also susceptible to attacks against the safety of the people, if the actuators are faulty or being hacked [6].

In this respect, there is increasing attention lately towards designing and developing fully secure and privacy preserving IoT systems. The main requirements for secure IoT systems are: (i) to exchange information from the devices to the applications in a secure way, (ii) to safeguard users' and citizens' private information, and (iii) to provide reliable information. To meet these requirements, IoT systems have to include from their design phase functionalities for secure device configuration, encryption, confidentiality, device and user authentication and access control, integrity protection, data minimization, etc. All these functionalities have to be included in the design phase of the IoT systems, because any post-mortem corrections will only cover some holes but won't provide full-scale security [7].

In the previous two versions of the IERC book [8, 9], we have extensively covered the areas of security and privacy in IoT. In this chapter we will focus on another very important area for ensuring the provision of reliable information and for maximizing the security, privacy and safety of IoT: "Trust". The remainder of this chapter is structured as follows: in Section 6.2 the basic concepts of trust in the IoT are described, together with the reasons for evaluating trust in the IoT world. In Section 6.3 we provide the basic concepts of trust management in the IoT, while in Sections 6.4 and 6.5 we present ways to calculate the trustworthiness of IoT devices and services. In Section 6.6 we present an analysis of using Trust with regards to privacy and personal data sharing. In Section 6.7 we present the improvement of the authorization mechanism with the usage of trust and reputation. In Sections 6.8 and 6.9 we present two examples of use of trust evaluation for an indoor positioning system and for improving the routing mechanism for increased confidentiality. Section 6.10 concludes the chapter.

6.2 The Need for Evaluating Trust in IoT

Trust is a very important concept for IoT since it can affect the adoption of the IoT systems by the humans. It is reasonable to assume that if the humans do not trust an IoT system and its components, they will not be willing to use it. The same stands for the service providers, the municipalities, the companies, and all kinds of IoT stakeholders. If they are not convinced that the IoT systems are reliable, they will not be willing to invest in them. Trust is closely interconnected with reliability and reputation. In Information and Communication Technologies (ICT) the concept of trust has been considered crucial for any digital interaction between multiple entities.

The concept of Trust can be defined as the level of confidence that an entity has on another entity to behave certainly in a given situation [10]. Up to recently, the notion of Trust was only used for humans, but lately it is also associated with machines, devices and software. Here, we also have to make a distinction between Trust and Trustworthiness. Trust can be considered as subjective, because it is a belief of an entity (user, device, etc.) that another entity is functioning according to some predefined criteria, and these criteria are subjective to the former entity. On the contrary, Trustworthiness, which is an abstract concept, is considered as objective, because it is described as a metric of how much an entity deserves the trust of other entities [11]. This metric is built upon several criteria, i.e. evidence of current and past behaviour, availability, data reliability, security, etc. Trustworthiness can also be calculated as "absolute" or as "relative" to other entities. For example, we can say that a device is trustworthy in general or that it is more trustworthy than its neighbours [12].

Another very important concept is the "Reputation" of an entity [13]. Although sometimes it is used interchangeably with trustworthiness, reputation is considered as an estimator of the trustworthiness of an entity according to the criteria of another entity. Since the trustworthiness of an entity is very difficult to be evaluated, the metric that is widely used instead is the reputation. In order to calculate the reputation of an entity, the metrics of multiple other entities are fused and compared according to certain criteria.

As mentioned above, IoT systems have to be trustworthy so that they are adopted by all stakeholders. The trust in the IoT domain can be considered at many scenarios which include information exchange between the various entities of the system. Since users and devices are exchanging information between each other, we can consider trust (i) from users to devices that send them information, (ii) from devices to users that send actuating commands, and (iii) between devices that exchange information and actuating commands.

For example, in Machine-to-Machine type communications (M2M) the devices that are exchanging information have to know the reputation of other devices so that only the devices that are trustworthy will handle sensitive or critical information. So, in a scenario where a temperature sensor sends commands to the air-conditioning system to turn on the heating because the temperature in the room is very low, the air-conditioning should be sure that the temperature sensor is trustworthy in order to execute the command.

Furthermore, only trusted users have to be allowed to manage critical data or actuators. This is quite important, because in a scenario involving controlling of physical objects, e.g. doors, windows or even fire-extinguishers, malicious (untrusted) users may create incidents against the safety of other users. However, since in the IoT devices are also able to control other devices, these incidents can also occur not only by user actions (i.e. hacking devices), but also by faulty or malfunctioning devices.

Another scenario can be assumed when users are receiving measurements from devices, i.e. measurements for traffic in the center of the city in order to identify the fastest route towards their work. If the system does not provide them reliable traffic information, the users will stop using this system, because they will not trust it.

Apart from the previous examples that are mostly related to providing reliable applications and services, trust in the IoT can also be related to the reliable configuration of the various system components. One such example will be given in Section 9.6. Trust can be included in all types of cooperative networking mechanisms, for example as described in [14] in cooperative spectrum sensing and assignment, where measurements from various devices are fused in a gateway for identifying spectrum opportunities and for deciding which is the best spectrum portion to operate on. Any measurements from untrusted sources may affect the decision of the system and may result in degraded system performance.

It is evident that considering Trust in the design of an IoT system is of outmost importance for improving its reliability, its security and the safety of its users. In the next sections we will discuss the recent approaches within IERC for evaluating and managing trust in IoT systems.

6.3 Trust Management in IoT

The main objective of a Trust management system in IoT is to be able to evaluate the trustworthiness of various components of the system and to use these values in order to provide reputation information to users of the IoT services or to internal configuration services.

Trust management systems use trust and reputation models that are based on five generic steps, as described in [15] and also discussed in [12]. The main goal is to enable one entity (human or device user or a service) to identify the entity or group of entities that are more trustworthy for a certain transaction, based on specific criteria. As described in [16] any IoT trust model should be designed according to the following:

- **Observation**: This step is the most important step since it is responsible for monitoring the parameters of the system entities and their behaviour, allowing the extraction of results with regards to the trustworthiness of the entities. The monitoring of these parameters can be performed by the system devices or by specific entities that are called observers. The collected information can originate from standalone observers or from groups of observers, which then fuse the information for extracting more objective results.
- **Scoring**: When the observers gather the information for an entity they can give it a proper weight which will result in a reputation score. This will be done for all entities in the system (considering that adequate information has been gathered). This reputation score can be given by an interested agent or a centralized entity or by many entities collectively. Finally, the reputation scores can be used in order to rank the entities in terms of trustworthiness according to some criteria.
- **Selection**: Once the reputation scores and ranks are in place, the next step is to select the entity which is more appropriate for a specific transaction, i.e. that provides a specific IoT service. Of course this service might be provided by more than one entity, thus the user has to select the most appropriate according to some criteria.

- **Transaction**: When a service has been selected, the transaction takes place and more information regarding the entity (that provides the service) is being gathered by the system components, as a feedback.
- **Rewarding and punishing entities**: Trust management systems should also include functionalities for rewarding the entities that are performing according to the criteria and have high reputation. At the same time, the system must punish malicious or untrustworthy entities that may negatively affect system decisions or the systems' overall reliability and trustworthiness.

Based on the above and as described in [16], a trust model for the IoT can be split in two main sub-models: (i) a trust evaluation model and (ii) a reaction model. The Trust evaluation model is responsible for gathering trust metrics and trust ratings for the system entities and evaluating them for extracting their reputation, while the reaction model is responsible for reacting to these reputation evaluations, either by rewarding or by punishing the entities.

The trust evaluation model has to be lightweight, keeping a small state that is updated regularly, so that it can also run on constrained devices. For the trust evaluation model, as proposed in RERUM [16], the main idea is that there is a set of observers that are providing trust ratings for a specific entity in mind (be it software, hardware, user or object). These trust ratings are trust values that relate to the confidence that this observer has on this entity according to some criteria. Trust ratings can also be provided by the administrator of the system or by other users that have had past interactions with this entity. These trust ratings are then fused into a centralized component (i.e. reputation manager) that extracts the reputation of this entity. Then, when a user, a service or another entity wants to interact with the entity under evaluation, it queries this centralised component to get the reputation and decide according to its own rules if it can trust this entity or not.

A reacting model can be considered as another set of rules that describe the actions of the system when a reputation for an entity is evaluated. Any reputation change may trigger reactions by the system [16]. For example, when a reputation of a trusted entity is being decreased, an alert may be triggered so that the system will search to find what is the cause of this reputation decrease. On the contrary, when a reputation of an untrusted entity is increased, another alert may be triggered so that this entity will be closely monitored to identify if it has become trusted or not. The reactions are based on specific rules that are mainly being defined by the system administrator. Various reactions can be defined, i.e. logging alerts, warning administrators, disabling or re-enabling services, stopping/starting gathering data from devices, initiating networking or system configuration mechanisms, warning users, etc.

In the following subsections, we present details for the trust evaluation model, as described in the RERUM [17] and Sociotal [18] projects. The focus is on devices and services, which are of outmost importance for the IoT. Although the end users are also very important when interacting with IoT systems, the trust evaluation for users is not discussed within this chapter since existing schemes for user reputation in the Internet can be applied [17, 20, 21].

6.4 Trust for Devices

The trust model for IoT devices aims to improve the reliability and trust-worthiness in IoT scenarios where disparate and unknown devices interact each other and provide data to IoT applications. The device-based trust model follows a multidimensional or multi-layered approach to calculate the overall trustworthiness of an IoT device. The model describes the procedure employed to quantify several trust dimensions (or trust metrics). Then, the dimension's values are aggregated to come up with a final score of trust i.e. by means of fuzzy logic or data fusion techniques such as the Dempster Shafer theory of evidence to avoid outliers or malicious nodes [22].

The trust dimensions correspond to different properties that have to be taken into account in the IoT paradigm. Contrary to past approaches that considered only reputation between different devices and data reliability, lately other parameters such as communication reliability, security aspects and social relationships between the devices are being considered. In the end, this approach leads to a more accurate and reliable value of trustworthiness about a given IoT device, which can be exploited either for improving the reliability of the provided services or for increasing the overall security of the IoT system.

The trust model follows a hierarchical and a layered approach in which the different dimensions are split in categories and subcategories, which in turn are composed by measurable properties. This hierarchical approach enables the trust model to be extensible, allowing users to consider and include new properties to the model. Nonetheless, the trust quantification procedure is the same regardless of the amount of properties taken into account. In fact, some of the trust properties explained below could be optional in case the implementation of the IoT system is unable to measure these properties. Of course in that case the resulting trustworthiness value of the device will be sub-optimal, but it will give a good indication [23].

The trust dimensions can be measured in different layers within an IoT network. Some of them can be measured on the devices themselves and the values will be exchanged between the devices and fused in order to extract the reputation of each of their neighbour devices. Other dimensions may be calculated at cluster heads or gateways, which will do the fusion of the reports of the devices and then they will feed back the results to the devices. This approach may save enough computational resources on the devices in case the trustworthiness evaluation is complex.

Finally, some dimensions may also be calculated at the backbone cloud servers or the IoT middleware, where centralized trust management schemes may be employed, which will allow the fusion of measurements from more devices connected to different gateways to have a more accurate reputation evaluation for the devices.

In IoT the evaluation of the trustworthiness of a device can be generally based on multiple criteria that can be grouped into 5 categories: (i) communication criteria, (ii) security criteria, (iii) data-based criteria, (iv) social relationship criteria, and (v) reputation criteria.

6.4.1 Communication-based Trust

The communication based criteria correspond to the quality of the communication links between the devices. Although someone may think that the communication link quality is not directly related with the trustworthiness of the devices, this is not entirely true because the link quality may affect significantly the performance of the device's transmissions. This will in turn affect the Quality of Service provided by this device (in terms of throughput, delay, jitter, etc.) and its availability.

Within RERUM, the communication based trust criteria are mainly used for evaluating the networking related trustworthiness of the devices which is then used to consider the trusted devices within network-related cooperative mechanisms such as cooperative routing, spectrum or channel allocation, network monitoring, etc. In this respect, the main criterion considered is the link quality statistics based on a link quality metric. In RERUM, the chosen metric is the *Expected Transmission Count* (ETX) metric which has been proved in the literature that is quite accurate in evaluating the reliability of the link between any two nodes.

The ETX is very widely used for routing mechanisms because apart from providing good reliability results it is also quite simple and computationally efficient, so that it can be easily calculated even in the very constrained IoT devices.

As described in [25], the ETX calculated for node *i* by node *j* is defined as:

$$ETX_{i,j} = \frac{1}{f_{i,j} \cdot r_{i,j}},$$

where $f_{i,j}$ is the metric for the forward delivery ratio, namely the probability that a packet sent from node *i* is received by node *j*, and $r_{i,j}$ is the reverse delivery ratio, namely the probability that the acknowledgement packet from node *j* will be received by node i.

It is easily anticipated that the ETX is a metric of the retransmissions that a device is performing in order to successfully transmit a packet to the destination.

Basically, the ETX expresses the average number of transmissions that are required for a successful delivery of a packet to its destination when there are transmission failures due to degradation of link quality (e.g. interference, collisions, etc.).

6.4.2 Security-based Trust

The security trust criteria are mainly related with the behaviour of a device as this is anticipated by its neighbours. In the literature, these criteria are also described as behavioural trust metrics.

These metrics correspond to specific types of attacks as described in [26] and presented in Table 6.1.

By evaluating and fusing these metrics, the security-based trust of the devices can be calculated, which will show how susceptible this device is in these types of attacks, affecting it overall trustworthiness and the way the rest of the neighbours behave towards this device.

These metrics can be calculated mainly at the device level or at the cluster head/gateway level, when the devices are incapable (in terms of resources) to do these calculations. In order to calculate these metrics at the device level, the devices have to be able to go into promiscuous mode [16].

If one wants to measure some of the metrics of the table (i.e. data/control packets forwarded, metric No. 1 in the table), every time a device sends a packet to one of its neighbours (in a multihop network) it should enter into promiscuous mode so that it monitors if the destination neighbour forwards the

Table 6.1 Neighbour behaviour monitoring [26]

No	Trust Metric	Monitored Behaviour	Attack Addressed
1	Data/control packets forwarded	Data/control message/packet forwarding	Black-hole, sinkhole, selective forwarding, denial of service, selfish behaviour, Control/routing message dropping
2	Data/control packet precision	Data integrity	Data message modification, Sybil and any attack based on routing protocol message modification
3	Availability based on beacon/hello messages	Timely transmission of periodic routing information reporting link/node availability	Passive eavesdropping, selfish node
4	Packet address modified	Address of forwarded packets	Sybil, wormhole
5	Cryptography	Capability to perform encryption	Authentication attacks
6	Routing protocol execution	Routing protocol specific actions (reaction to specific routing messages)	Misbehaviours related to specific routing protocol actions
7	Battery/lifetime	Remaining power resources	Node availability
8	Sensing communication	Reporting of events (application specific)	Selfish node behaviour at application level

correct packet, if it forwards a modified packet or if it drops the packet. Then, it can change the respective trust rating for this neighbour device accordingly.

In RERUM's view, it can be assumed that the metrics (1), (2), (3), (4), (5) and (8) are the most important ones, while the others can be used in specific cases.

The metrics can be used either "as is" or by assigning different weights to each one for giving larger weight values to the most "important" metrics according to the application what will use the trustworthiness value of the device. One such example is given in [26]:

$$BR_{i,j} = \sum w^s \times BC^s_{i,j}, \text{with} \sum_{s=1}^{n} w^s = 1.$$

In RERUM [27], we have used formulas for the metrics No. 1 and No. 2 in the table above, namely for packet delivery and packet integrity. These are assumed to be the most commonly used in this type of trust criteria because they represent the most common attacks for malicious users in IoT networks.

All devices within an IoT network are assumed to be monitoring the behaviour of their neighbours when they are interacting with them.

For these two metrics, the following statistics can be used: (i) *Packet Drop Rate* (PDR), as the ratio of the number of dropped packets divided by the total number of received packets and (ii) *Packet Modification Rate* (PMR), as the ratio of the number of modified packets divided by the total number of forwarded packets.

However, these metrics correspond to the values observed by one device for one of its neighbours. Assume that a receiver device 'j' receives a packet, each neighbour 'i' overhears the forwarding behaviour of 'j' and updates accordingly the values of $PDR_{i,j}$ and $PMR_{i,j}$. Then, we can use a combined metric called aggregate *Misbehaviour Rate* (MBR) for the device 'j' as perceived by device i is calculated as:

$$MBR_{i,j} = w \times PDR_{i,j} + (1 - w) \times PMR_{i,j}$$

where $w \in [0,1]$ is a user-defined weight controlling the balance between the behavioural statistics.

6.4.3 Data-Reliability based Trust

One of the most important trust metrics for IoT devices is related to the reliability of the data they produce. By using the term "data" we refer to the measurements the IoT devices are producing from their onboard sensors, i.e. environmental, location, energy, etc. These measurements are being used by the services of the system and if they are unreliable they may severely degrade the trustworthiness of the overall system. Consider for example a weather station producing wrong values for the temperature and the rain level in the centre of the city and the citizens are falsely informed and are not properly dressed. Another example may be regarding the alerts for fire or hazardous gases. It is evident thus that the data reliability is very important because it can even affect the safety of the users/citizens.

The data reliability based trust metric is also described in the literature as "service-based metric" [16]. Its evaluation is done by comparing the measurements with known measurement patterns, past measurements or measurements of other devices at the same area, monitoring the same physical object and the same property of the object. This means that we should only compare temperature measurements from different sensors monitoring the same room and not different rooms or measurements of temperature with humidity. The goal is to identify inaccuracies in the measurements observed by the devices. In this direction, a statistical analysis of the measurements' time series can be done (i.e. the deviation from the average value reported in X previous timeslots) and/or compared to the measurements reported by another similar device. For this type of calculation, there have been proposed many techniques in the literature for i.e. outlier detection in WSNs, see the references in [28].

What is different in the IoT world, as described in RERUM, is that the IoT devices may have various sensors of different types onboard and may be providing multiple services. As a result, when the system needs to evaluate the data-based trust metric, this evaluation must be done separately for each one of these services and then it can be combined, if needed, to calculate the overall data-based trustworthiness of the device. In most cases, the applications or the functions that will use the data-based trust rating will only need the rating for one service and the overall trust rating may not be of importance for them. However, for self-monitoring purposes, the overall trust rating might also be important.

Let's assume that each device can provide 'N' services, then N data-based trust ratings, one for each service it provides can be calculated. A low trust rating for one service does not mean that other services provided by different sensors will also be unreliable. However, combining the trust ratings for services provided by a specific sensor can provide results about the malfunctioning of that sensor or its driver being hacked. Furthermore, the fusion of the trust ratings of all services can only give a hint if the node is tampered with/hacked so that it reports intentionally false measurements.

So, we can have trust metrics as below:

$$OSTM_i = \sum_{S_x=1}^{N} w^{S_x} \times STM^{S_x}i,$$

where OSTM is the overall service based trust metric and the STM is the trust metric for each one of the provided services Sx.

6.4.4 Social Relationship based Trust

In IoT, social parameters can also be used for evaluating the trust rating of IoT devices. These social parameters are based on the emerging Social IoT (SioT) paradigm, which assumes that devices can establish social relationships with each other. In such a case, devices are assumed to be grouped into trust bubbles or communities based on their social relationships, i.e. if they belong to the same owner, if their owners are friends, are working together, if they are located at the same area, if they have the same manufacturer, etc. The Community of the devices is formed when the devices that share common interests are interacting and the more they interact the stronger becomes the trust relationship between them [23].

An IoT trust model has to consider the social relationship between a device 'i' when assessing the trust of a device 'j'. Different weights can be given to the relationship between the devices considering the links among them. The weights assigned by the trust model to the social relationships should be configurable by the user in the interval [0..1] and should satisfy:

$$W_{B_p} > W_{B_f} > W_{B_o} > W_C$$

Where B_p is the Personal Bubble, B_f is the Family Bubble, B_o is the Owner Bubble and C is the Community Bubble. Apart from this, when the devices do not belog in one of these bubbles, the trust model can calculate the degree of *Interest-In-Common* or the *Friends-In-Common*. The Interest-In-Common I_j^i can be calculated as the ratio between the interest that both devices share over the total amount of interests of the evaluator device

$$I_j^i = \frac{interest(i) \cap interest(j)}{interest(i)}.$$

Similarly, the Friends-In-Common F_j^i can be calculated as the ratio between the number of friends that both devices have in common, and the total amount of friends of the evaluator device

$$F_j^i = \frac{friends(i) \cap friends(j)}{friends(i)}.$$

It should be noticed that to quantify the interests and friends in common the devices should be able to exchange, in a common way, their list of interests (e.g. services and capabilities) as well as the lists of friends.

6.4.5 Reputation based Trust

As mentioned before, an IoT trust model should consider recommendations from multiple devices about a particular device j. Let O_j^i be the Opinion score about device j given by device i. It is also reasonable to assume that the opinions of different devices may have different impact on the opinion score of other devices, that's why there can be weights for each one of the "recommender" devices. This weight can be calculated based on the past behaviour of this device in the opinion scores or also on the trustworthiness of the device [23]. Thus, the opinions are subject to a credibility process where each reputation evidence coming from a device i is subject to credibility factor Cr_i in the interval $[0..1]$, where 1 represents the highest credibility. Therefore, the Reputation property in our trust model is given by $R_j^i = O_j^i * Cr_i$.

Since the opinion scores are calculated by the trust ratings provided by the devices for their neighbours, the results can be biased leading to uncertainty. For this type of reputation evaluation, other techniques for trust fusion can be used, i.e. the Dempster Shafer theory of evidence, which is a powerful mathematical framework able to handle uncertainty of the complete probabilistic model describing the system under consideration.

The calculation of the reputation metric can be done either at the device level, at the gateways, or even at a centralized or cloud based IoT middleware [16]. If calculated at the device level, each device should store the direct evidences and recommendations provided by other devices to quantify trust of each neighbour. However, this can be quite demanding in terms of computational and storage resources and might not be appropriate for constrained IoT devices. Thus, either evidences about devices which they do not interact for a long period of time should be discarded or the evaluation of the reputation trust should be escalated to the cluster heads, gateways or the middleware.

6.5 Trust for IoT Services

The IoT Services provide streams of information towards the end users. Thus when evaluating the reputation of a service, the goal is to provide enough information so that a user can have an answer to the question if he can rely on a specific service or if the service provides reliable measurements. As mentioned in the beginning of Section 9.3, a user has to query the reputation manager for getting the reputation value of that service. We can assume that for privacy reasons only authorized users are allowed to query the reputation manager for specific services.

IoT services can be either simple services provided by a single device or composed services that combine data from many devices. Of course, behind the provision of the service lies a business logic that also has some rules for managing the data from the devices. The reputation of a service is directly related with the reliability of the data of this service. As a result, for evaluating the reputation of a service, the trust rating of its data stream has to be evaluated. Thus, an observer has to be allocated to monitor this data stream. The observer should basically extract statistics for the data stream, in order to be able to identify changes in the pattern of the data stream, i.e. to identify when there are jumps or values that are outside the "normal" limits of the data stream [16].

For the statistics of the data stream, the first calculation that has to be done is the average value, that can be calculated as an overall average or as a moving average on a sliding window (according to the criteria of the administrator and the properties of the data stream). Here the challenge is to be able to calculate the average without using too much storage, so that even constrained devices will be able to calculate it. Then, the observer has to calculate also the limits and the thresholds of the data stream (in terms of minimum and maximum value) so that an alert will be fired if a value outside these thresholds is measured [16]. For example, when measuring the temperature in a room, it might have been noticed that in the past the minimum value was around 5 degrees and the maximum around 35 degrees. If the temperature monitoring service provides values of around 50 degrees, an alert has to be fired for a possible fire in the apartment or for a possible tampering with the service's data (i.e. a hacked device or a n intermediate entity altering the measurements). In the latter case, the reputation of the service has to be lowered.

Apart from the values outside the thresholds, sudden jumps in the data stream might cause change in the reputation of the service [16]. For example, in the previous scenario of a temperature monitoring service, if the current temperature of the room is around 10 degrees and suddenly the service starts providing values around 25 degrees this might fire an alarm despite the fact that the values are within the thresholds. Such a sudden jump has to be evaluated because it might mean that the service might be providing false values and its reputation has to be decreased. For this reason, the alarm might to be a warning to the administrator of the system to check what is happening in that room. Another type of an alert, may cause the cross-evaluation of the values of the temperature service with the values of other services, i.e. of a smoke detector service to see if there is indeed a fire in that room, etc.

It can be easily understood from the latter scenario that in order to evaluate the reputation of a service, the calculation of the statistics of its data stream might not be enough. Thus, there needs to be a mechanism to allow the cross-evaluation of the statistics of different data streams (assuming that some data streams are known to be trustworthy).

For the usage of the statistics of the data streams, the definition of the thresholds and the identification of the jumps, specific rules have to be defined either by the administrator of the system or by the users that want to receive a service [16]. Within RERUM, the expert system CLIPS [29] has been selected for the implementation of the rules in a simple but powerful way.

6.6 Consent and Trust in Personal Data Sharing

The volume of data is doubling every two years, of which two thirds is generated by individuals, in particular with adoption of new wearable devices [30]. This growth has been driven by the increasing of both the number of connected devices in our lives as well as their capabilities. This trend looks set to continue with data traffic from IoT devices rising from 2% share of the total in 2013 to 17% in 2020. Only considering the Public Health sector, sharing of personal data is estimated to generate 100Bn EUR value per year. This derives from the creation of new services such as those providing holistic approach to healthcare, where prevention and caring of long-term conditions can be made more effective by combining information beyond those included only in medical records, but including also any related life style information (such as shopping and dietary habits, fitness/exercise information etc).

In the current IoT service model, personal data are mostly collected by a multiplicity of Service Providers, each one offering a dedicated service, most of the time provided through a freemium model [31], whose main revenues stream is generated by third party exploitation of generated data for target advertisement.

This currently undermines individuals' trust in sharing IoT personal data, thus hindering its associated value. A recent Digital Catapult report on Personal Data and Trust [32] highlighted that 60% of consumers are uncomfortable about sharing their data, with a further 14% feeling so uncomfortable that they do not want to share their data at all. Individuals' reluctance to share personal data becomes higher when commercial purpose is foreseen while more confidence is put in sharing data for research purposes. However, people feel uncomfortable with their information being used for secondary purposes if not enough trust is put in the organization originally collecting the data and re-distributing them to third parties [33]. Preparedness of individuals to share their data varies considerably by sector, with more than a third of individuals trusting banks and the public sector, but less than 5% trusting mobile network operators, utilities, retail and media companies. In general 80% of consumers feel organisations hold their data solely for economic gain. Even for public sector organisations, only 45% of consumers believe they hold data for their benefit.

There is a need to regain individuals' trust by increasing transparency on how data are collected, managed and shared. *Control* is the key and to support this change in the current trend, the new General Data Protection Regulation [34] (aka GDPR), recently approved and in force by early 2018, is putting the end-user (aka *the individual*) at the center of this process, while promising expensive sanctions for those businesses big and small failing to comply to its principles (e.g., up to 100 Mio or 4% of their annual turnover fines for big corporates and up to 100K or 2% of annual turnover for SMEs).

Figure 6.1 shows what are the elements required to develop a personal data sharing ecosystem, where trust should be maintained by giving individual control on how their personal data are collected and further used.

Attribute Providers collect personal data through the provisioning of a service as part of their day-to-day operations (e.g., banks, utility suppliers, IoT service providers, etc.). To avoid to lock such data in silo'ed systems, and to allow further access, reuse and combination of them for creation of new services by a growing ecosystem of SMEs, data need to be brokered according to well-defined rules (aka *the Scheme*), enforced by *certified* Scheme Operators.

For ensuring compliance to GDRP, while increasing individuals' trust, the envisioned Scheme should set, among others, the following principles:

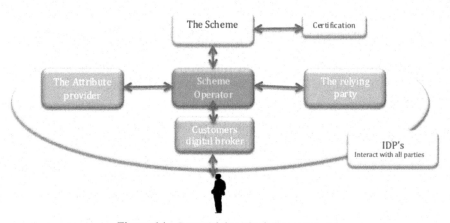

Figure 6.1 Personal data sharing ecosystem.

- *Transparency*: Privacy Notices for data sharing should be easy to access and to understand, explaining how data are processed, what are the individuals' rights and how they can be enforced;
- *Consent*: Valid consent must be explicit for data collected and the purpose for the data collection should be stated. Data controllers must be able to collect "consent" form end-users (opt-in) and consent might be withdrawn;
- *Erasure*: Attribute Providers (e.g., the data controllers) are the entry point for the erasure requests and need to inform third parties (e.g., the Relying Parties).

If control means trust for individuals [35], to exercise this control, hence the consent to sharing cross-domain personal data, there is the need for tools and open standards. Consent Receipt [36] represents one of such tools.

The Consent Receipt inherently, *by being a record of consent given at the point of consent (e.g., when first accessing a service)*, provides proof of consent and delivers contact information to communicate about consent directly to the end user. According to GDPR and in order to guarantee individuals' trust, consent should be: *freely given* (opt-in); *informed*, i.e., 'no legitimate interest' in using collected data should be allowed; *specific*, i.e., bound to the purpose the data are collected for; *unambiguous* and *transparent*, i.e., additional personal data cannot be vaguely collected while offering a service; *dynamic*, i.e., it can change over time and be revoked at any time.

The Consent Receipt provides the evidence that the consent for personal data sharing is properly collected and guarantee individual control over it, thus maintaining trust in the created ecosystem.

Figure 6.2 provides a summary of a Minimum Viable Consent Receipt standard's elements, currently under development by the Kantara Initiative through its Consent and Information Sharing Working Group and the support of the Digital Catapult Personal Data Network (https://pdtn.org).

In particular:

- *Header*: The purpose of which is to set out administrative fields for the consent transaction, including a unique Consent ID;
- *PI (Personal Information) Controller Information*: This section identifies the individual and company that is accountable for data protection and the privacy policy (included in the receipt or linked to otherwise) to which the consent is bound;
- *Purpose Specification*: This section clearly specifies the purpose(s) for which Controller is collecting additional Personally Identifiable Information [37];

Figure 6.2 Consent receipt structure.

- *Personally Identifiable Information*: This section specifies the personal information categories and related attribute collect by the PI Controller;
- *Information Sharing*: When applicable and stated in the Privacy Policy, the purpose of this section is to provide the individual with information about how their information is shared with third parties;
- *Scope*: This section specifies the technical and policy scope within which the collected data are used.

Like a paper receipt for any purchased good, it is clear how issuing end users with Consent Receipts, adequately certified by a Scheme Operator, for each digital service developed by a Service Provider using personal data collected by Attribute Providers, gives them a trusted tool to clearly understand how their data are used and to control how they are eventually shared. The same tool allows also to easily revoking access to such data with possibility to backward notify all the third parties accessing the same data, thus guaranteeing *the right for erasure*.

To ensure use of such trust tool, some additional elements are requested to create the Trust Framework encapsulated in the "Scheme" overarching the personal data sharing and operationalized by the certified Scheme Operators. Figure 6.3 shows the elements of the so defined Open Consent Framework [38].

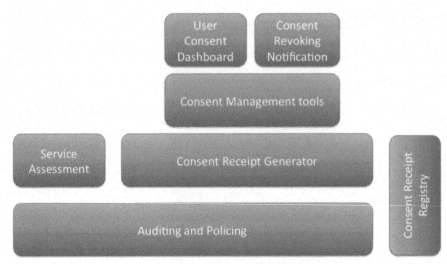

Figure 6.3   Open consent framework.

First of all, certified authorities perform a Service Assessment of Relying Parties that develop services using personal data, in order to provide a *data protection impact assessment* and to collect the information required to pre-fill the Consent Receipt fields, specifying what data and how they are collected, used and shared according to stated privacy policies. The result of such assessment is used to pre-configure a Consent Receipt Generator, the access to which is provided to the given Relying Party as Service (Consent Receipt As A Service, CRaaS). Unique Consent Receipt IDs, useful for auditing purposes, are created by the Scheme Operator and assigned to each generated receipt. Along with the Consent Receipt ID, the remaining Consent Receipt fields are filled at run time.

A first implementation and the related open APIs of a Consent Receipt Generator can be found at: http://api.consentreceipt.org. For easiness of management a JSON Web Token conversion of a Consent Receipt generated by the Consent Receipt Generator is returned.

With this minimum set of services in place, third parties can develop Auditing and Policing functionalities (e.g., similar to EuroPriSe [39] is doing for website) aiming to verify that data are processed and used by Relying Parties according to what stated in the given Consent Receipt. The result of such auditing can enforce policing actions towards organization failing to comply with the agreed principles and to build a Consent Receipt Registry providing a transparent Kitemark [40] of *compliant organizations*. This will allow end-users to monitor *reputation* of the organizations they give consent to access to their own data.

On the other end, a set of end-users facing tools allow, among others, individuals to manage consent, collect and group receipts, as well as visualize and track shared personal data, through a *User Consent Dashboard* [41]. Currently more UX research is undergoing in order to understand, from an end-user perspective, how to better visualize in the Consent Receipt and associated Dashboard information about the type of data collected and how they are used. The British Standard Institute (BSI) and Digital Catapult are currently developing a new Publicly Available Specification (PAS) [42] defining a number of icons providing such information, using traffic light colour codes similar to those used to classify food composition [43].

To support Consent revocation, achieved by handing over a given receipt to the Relying Party providing the subscribed service, and to notify involved third parties to remove collected data, an additional set of Consent Revoking Notification tools need to be developed.

By achieving end users trust through the above presented Open Consent Framework, the Personal Data Sharing of IoT Services ecosystem (Figure 6.4)

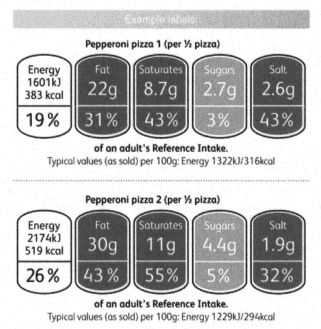

Figure 6.4 Example of food labels inspiring the data labels.

can be further grown with the future development and deployment of Customer Digital Agent (CDA), e.g., organizations, autonomous agents, robo advisors, or ultimately blockchain-based smart contracts (https://www.ethereum.org) that offer and manage service subscription requests on behalf of end-users and based on context and on user preferences as learned by previously accepted services and their issued receipts. This will open up potential for a new personal data market for IoT services, where data are shared with individual trust.

6.7 Using Trust in Authorization

The IoT Access control system can implement a Trust Model in order to enable secure and reliable interactions between granted and trustworthy entities. This mechanism can be deployed on IoT scenarios where smart objects can maintain social relationships, composing different kinds of groups of entities called "bubbles" (e.g. Personal, Family, Office or Community). According to Figure 6.5, each bubble is made up of a set of smart objects, along with an Authorization Manager, which is responsible for generating authorization credentials for smart objects. Furthermore, each smart object have a Trust Manager, which is in charge of assessing the trustworthiness of the other involved entities [23].

The main entities involved in the trust-based access control process are the following:

- **Smart object**. It is a device (e.g. a smartphone, printer, camera, sensor, etc.) that can act both as a CoAP client and a CoAP server offering services (e.g. temperature, location, etc.) in an IoT environment.

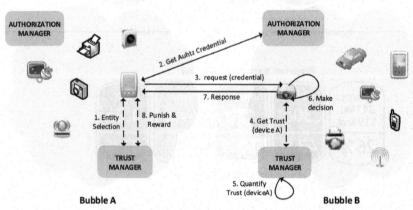

Figure 6.5 Sample scenario for Trust-based authorization in IoT.

- **Trust Manager**. It is the component implementing the proposed trust model. In the case of a smart object with tight resource constraints (i.e., class 0 or class 1 device), the Trust Manager is deployed as separate network element, such as a gateway. In the case of more powerful smart objects (at least class 2 devices), the Trust Manager is a part of the devices.
- **Authorization Manager**. It is responsible for generating and sending authorization tokens to smart objects. Additionally, it is composed of two subcomponents; the Policy Decision Point (PDP), which is in charge of making authorization decisions based on a XACML engine, and the Token Manager, which generates authorization credentials according to the authorization decisions.

In a trust-aware access control system, an **intra-bubble** communication happens when a smart object attempts to access another smart object that is part of the same bubble. Figure 6.5 shows the interactions at high level in the case of an **inter-bubble** communication between smart objects from different bubbles. Under this scenario, the purpose of the Trust Manager (TM) is twofold. On the one hand, it is used by the **requester** smart object to know the most trustworthy target among a set of devices providing the same service. On the other hand, it is employed by the **target** smart object in order to get the trust value associated with the requester under a specific transaction. This value is used, along with the authorization credential that is previously obtained from the Authorization Manager, in order to make the access control decision.

The process carried out during the trust-based access control, depicted in Figure 6.5, is summarized as follows. Firstly, the smart object A accesses its TM in order to know the most trustworthy smart object providing a specific resource in bubble B. The TM calculates the trust of the set of available devices in bubble B (a prior discovery of devices is assumed). Then, in step 2 device A obtains an authorization token for accessing to devices in B. The decision is made based on XACML policies evaluated by the policy engine. This stage is optional and it is supposed to be done not so often, as tokens are reusable. Afterwards, in step 3, the subject smart object A uses the authorization credential (authz token) for access to a service/resource being hosted on the target smart object B. The target acts as PEP (Policy Enforcement Point) that enforces the authorization rights defined in the token, taking into account as well actual context conditions. During this interaction the target device also considers the trust value associated to the requester device (i.e. smart object A). To this aim, it contacts its TM in bubble B, which quantifies in real time the trust based on previous evidences within A as well as actual conditions.

Then, in step 6, device B verifies that the obtained trust value is greater than a threshold value, which was specified as a condition in the token. If that condition is fulfilled, the request is accepted and the service is provided to the smart object A. Finally, in steps 7 and 8 (reward stage), the smart object A sends to its TM evidences feedback about the reliability of the interaction, in order to update the trust value associated to the smart object B, which is useful for future interactions.

6.8 Using Trust in an Indoor Positioning Solution

Smart buildings are comprised of devices integrated into the Internet infrastructure with network and processing abilities, which make them vulnerable to attacks and abuse. The associated services and resources can be accessed through mobile devices anytime and anywhere by common users, which may interact each other according to their levels of trust and reputation. In the smart buildings context, location-aware mechanisms for trust evaluation, can allow a user located at a certain room to share his data only with users located in his same location. In this way, a specific level of trust can be automatically established among people located in the same room, because all of them can be seen as belonging to the same ecosystem [44].

The effectiveness of location-aware security mechanisms is closely related to the accuracy of the location information and the definition of security zones, that is, the area where security aspects like access control, trust, reputation, etc. may be established. However, in the context of smart buildings, how this location information is obtained is a challenging task since traditional mechanisms such as GPS are not useful. The indoor localization mechanism for smart buildings is able to provide accurate location data to be included in security aspects of smart services. The proposed system is based on the use of sensors which are integrated in common smartphones built-in magnetic sensors to make security mechanisms totally independent on the type of devices and available signals in buildings. The sensed magnetic field is a combination of the effects of the Earth's magnetic field and that of surrounding objects. A methodological approach is used to generate the buildings maps containing the magnetic field distribution used as map of fingerprints. Then, based on such maps, location estimations are calculated using a combination of Radial Basis Functions Networks and Particles Filters [45].

The Access Control system can rely on the Indoor location enabler to make authorization decisions accordingly. In this way, devices can ask this service in order to get the distance where a requester user is placed when trying to access to their services; consequently, certain services can be only provided when users are placed inside the authorization zone of some smart objects. Figure 6.6 below depicts the proposed scenario to perform location-aware access control in indoor environments. The smartphone, acting as a subject, requests to get access a resource being provided by the target smart device. Before allowing it to access to his resource, the target device evaluates both the capability token as well as the subject's position, which must be located inside target's security zone. The context that determines the smart object B position comes from the indoor localization enabler.

Firstly, the use case requires an offline stage where the smartphone of user A contacts with the Authorization Manager in order to get an authorization credential to get access to smart objects. Notice that this phase requires the authentication process. Once the subject is successfully authenticated, the Authorization Manager evaluates the policies and generates (if allowed) a token with the set of privileges associated to the smart object. Then, the subject device wants to make use of a resource hosted by target device, and it uses the obtained token to present it against the target, which validates the token, see it the quantified trust value is over a threshold, and checks subject's position against a localization service, since only those devices located nearby are allowed to get access.

6.9 Using Trust in Routing

A different scenario for the application of trust management in IoT systems is related to improving the security, the privacy and the performance of a network of IoT devices. Assume that there is an IoT deployment with a

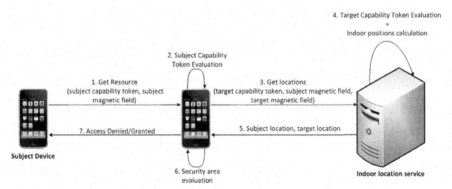

Figure 6.6 Location-aware access control for indoor environments.

large number of IoT devices that are forming a multi-hop sensor network. In such a network, the information from the leaf devices or any device has to pass through a number of intermediate devices before it reaches the gateway that will forward the measurements to the backbone middleware and the applications. If there are intermediate devices that are either tampered with, malicious or faulty, this may result to loss of information or to the provision of faulty/tampered information. Moreover, malicious devices may be able to get access to sensitive user information that is passed through them.

To avoid such scenarios, the evaluation of the trustworthiness of the devices can be used in the routing mechanism of the network, so that malicious or malfunctioning devices will be quickly identified and sensitive information to be passed only through trustworthy devices. As described in [27], the reputation evaluation of a network of IoT devices can be done very easily. Assuming that the devices are able to monitor the transmissions of their neighbours, the trust evaluation system can identify very quickly which devices are providing erroneous information. The main idea is that before the start of the trust evaluation all devices have a trust-rating of "unknown", which is then changed as the devices start to exchange data and observing the behaviour of their neighbour devices. Generally, the rules that can be applied are that the trust-belief for a device (i.e. how much we trust a device) should increase slowly, in order to be sure after many interactions that the device is trusted, but it should decrease faster, so that malicious or suspicious devices should be avoided.

When the reputations of the devices have been calculated, then these have to be included in the definition of the routing metric, to ensure that the nodes will be able to identify the routes to the gateway by avoiding suspicious or malicious devices. As shown in [46], including the device reputation in the routing mechanism can significantly improve the performance of the IoT network in terms of improved packet delivery ratio and throughput. This is justified because by avoiding malicious nodes, the percentage of packet losses or packet integrity fails will be minimized.

6.10 Conclusions

The IoT requires new adapted trust models able to overcome the drawbacks of traditional complex models that have not been tailored for the pervasive nature of such global ecosystem. The IoT trust management aims to improve the reliability and trustworthiness in IoT scenarios where disparate and unknown devices interact with each other. It is known that trust is closely inter-related with security and privacy. However, the inter-relationship is not purely bi-directional. If an entity is neither secure nor privacy preserving, then it should not be trusted. On the contrary, if an entity is secure and privacy preserving, this does not necessarily make it trustworthy for all users.

In this sense, this chapter has shown a trust model that follows a multidimensional approach to calculate the overall trustworthiness of an IoT device. It defines different criteria for the evaluation of the trustworthiness, such as communication, security, data-based criteria, social relationships, and reputation.

Moreover, the trust model provides means for detecting malfunctioning devices by checking if the provided values are inside a static range of values. To this aim, it relies on a rule based approach and fuzzy logic techniques for assessing the trustworthiness, which considers the plausibly, that is, whether the devices are generating correct values.

In addition, this chapter has shown the way the IoT trust management can leverage the access control by making authorization decisions based on quantified trust values as well as indoor localization context. In this sense, magnetic field techniques have shown its feasibility for providing accurate indoor localization positions with the aim of helping to make reliable authorization decisions.

Acknowledgment

This work is partially funded by the EU FP7 projects RERUM (GA no 609094), SOCIOTAL (GA no 609112) and UNIFY-IoT (GA no 688369).

Bibliography

[1] D. Evans. *The internet of things. How the Next Evolution of the Internet is Changing Everything*, Whitepaper, Cisco Internet Business Solutions Group (IBSG) 2011.

[2] A. Zanella, et al. Internet of things for smart cities. *Internet of Things Journal*, IEEE 1.1 (2014): 22–32.

[3] L. Gurgen, et al. Self-aware cyber-physical systems and applications in smart buildings and cities. *Proceedings of the Conference on Design, Automation and Test in Europe*. EDA Consortium, 2013.

[4] M. Weiser. *The computer for the 21st century*. Scientific american 265.3 (1991): 94–104.

[5] N. Petroulakis, et al. A lightweight framework for secure life-logging in smart environments. *Information Security Technical Report* 17.3 (2013): 58–70.

[6] E. Z. Tragos. et al. Enabling reliable and secure IoT-based smart city applications. *Proceedings of IEEE International Conference on Pervasive Computing and Communications Workshops* (PERCOM Workshops), 2014.

[7] H. C. Pohls, et al. RERUM: Building a reliable IoT upon privacy- and security-enabled smart objects. *Wireless Communications and Networking Conference Workshops* (WCNCW), 2014 IEEE. IEEE, 2014.

[8] O. Vermesan, and P. Friess (Eds). *Internet of Things-From research and innovation to Market Deployment.* River Publishers, 2014.

[9] O. Vermesan, and P. Friess (Eds). *Building the Hyperconnected Society: IoT Research and Innovation Value Chains, Ecosystems and Markets*, River Publishers, 2015.

[10] M. Ion, A. Danzi, H. Koshutanski, L. Telesca, A peer-to-peer multidi-mensional trust model for digital ecosystems, 2nd IEEE International Conference on Digital Ecosystems and Technologies, vol., no., pp. 461, 469, 26–29 Feb. 2008.

[11] G. Baldini, et. al., IoT Governance, Privacy and Security Issues, IERC whitepaper, 2015.

[12] N. Gruschka, D. Gessner (eds), Concepts and Solutions for Privacy and Security in the Resolution Infrastructure, IoT-A Deliverable D4.2, February 2012.

[13] Z. Yan, P. Zhang, and A. Vasilakos. A survey on trust management for Internet of Things. *Journal of Network and Computer Applications*, (42):120. March 2014.

[14] E. Z. Tragos, and V. Angelakis. Cognitive radio inspired m2m commu-nications. Proceedings of 16th International Symposium on Wireless Personal Multimedia Communications (WPMC), IEEE, 2013.

[15] S. Marti, H. Garcia-Molina, Taxonomy of trust: categorizing P2P reputation systems, *Computer Networks* 50 (4) pp. 472–484. Elsevier Science. Available via http://zoo.cs.yale.edu/classes/cs457/fall13/Tax onomyOfTrust.pdf. 2006

[16] D. Ruiz (Ed) et. al., Modelling the trustworthiness of the IoT, RERUM Deliverable D3.3, April 2016.

[17] RERUM, Reliable Resilient and Secure IoT for smart city applications, www.ict-rerum.eu.

[18] SOCIOTAL, Creating a socially aware and citizen-centric Internet of Things, www.sociotal.eu

[19] G. Zacharia, A. Moukas, and P. Maes. Collaborative reputation mechanisms for electronic marketplaces, *Decision Support Systems* 29.4 (2000): 371–388.

[20] J. D. Work, A. Blue, and R. Hoffman. Method and system for reputation evaluation of online users in a social networking scheme. U.S. Patent No. 8,010,460. 30 Aug. 2011.

[21] Z. Malik, and A. Bouguettaya. Reputation bootstrapping for trust establishment among web services. *Internet Computing*, IEEE 13.1 (2009): 40–47.

[22] A. P. Dempster, The Dempster-Shafer calculus for statisticians. *International Journal of Approximate Reasoning* 48.2 (2008): 365–377.

[23] P. N. Karamolegkos, et al. User-profile based communities assessment using clustering methods. Proceedings of 18th International Symposium on Personal, Indoor and Mobile Radio Communications, PIMRC IEEE, 2007.

[24] J. B. Bernabe, J. L. Hernandez Ramos, and A. F. Skarmeta Gomez. TACIoT: multidimensional trust-aware access control system for the Internet of Things. *Soft Computing* (2015): 1–17.

[25] S. Biswas, and R. Morris. ExOR: opportunistic multi-hop routing for wireless networks. *ACM SIGCOMM Computer Communication Review*. Vol. 35. No. 4. ACM, 2005.

[26] T. Zahariadis, et al. Trust management in wireless sensor networks. *European Transactions on Telecommunications* 21.4. pp. 386–395. 2010.

[27] E. Tragos (Ed) et. al., Introducing CR elements into smart objects towards enhanced interconnectivity for Smart City applications, RERUM Deliverable D4.1, March 2015.

[28] M. Gupta, J. Gao, C. Aggarwal, and J. Han. Outlier detection for temporal data. *Synthesis Lectures on Data Mining and Knowledge Discovery*, Vol. 5, No. 1, pp. 1–129. March 2014.

[29] CLIPS Reference Manual. CLIPS Basic Programming Guide version 6.30. (http://clipsrules.sourceforge.net/documentation/v630/bpg.pdf). last accessed: March 17th 2015.

[30] https://www.idc.com/getdoc.jsp?containerId=prUS40846515

[31] https://en.wikipedia.org/wiki/Freemium

[32] http://www.digitalcatapultcentre.org.uk/wp-content/uploads/2015/07/Trust-in-Personal-Data-A-UK-Review.pdf

[33] https://script-ed.org/article/share-and-share-alike-an-examination-of-trust-anonymisation-and-data-sharing-with-particular-reference-to-an-exploratory-research-project-investigating-attitudes-to-sharing-person al-data-with-the-pu/

[34] http://ec.europa.eu/justice/data-protection/

[35] https://kantarainitiative.org/confluence/display/infosharing/Home

[36] https://github.com/KantaraInitiative/CISWG/blob/master/MVCR-Spec /MVCR-v0.8/MVCR%20v0.7.9.md

[37] https://en.wikipedia.org/wiki/Personally_identifiable_information

[38] http://smartspecies.com/open-consent-framework/

[39] https://www.european-privacy-seal.eu/EPS-en/Home

[40] https://en.wikipedia.org/wiki/Kitemark

[41] https://eu-smartcities.eu/sites/all/files/Addressing%20Privacy%20in% 20Smart %20Cities%20-%20Chris%20Cooper%20-%20Consentua% 20API.pdf

[42] http://shop.bsigroup.com/navigate-by/pas/

[43] https://iapp.org/news/a/europes-privacy-seal-schemes-gradually-takin g-shape/

[44] M. Moreno, and A. F. Skarmeta. An indoor localization system based on 3D magnetic fingerprints for smart buildings. *Proceedings of International Conference on Computing & Communication Technologies-Research, Innovation, and Vision for the Future* (RIVF), RIVF IEEE, 2015.

[45] M. Nati, et. al., Device centric enablers for privacy and trust, SOCIO-TAL Deliverable D3.1.2, February 2015.

[46] E. Tragos, et. el., Improving the Trustworthiness of Ambient Assisted Living Applications, *Proceedings. Of WPMC* 2015, Hyderabad, India, 13–16 December 2015.

7

IoT Societal Impact – Legal Considerations and Perspectives

Arthur van der Wees, Janneke Breeuwsma and Andrea van Sleen

Arthur's Legal B.V., The Netherlands

7.1 The Relevance of Hyperconnectivity

Technology changes the world in a fast pace. Information, communication, internet and cloud computing technologies have shown and are showing this already on a daily basis, and have connected people, organisations and data. The Internet of Things (IoT) will push this process further, by hyperconnecting people, organisations and their data with billions of objects.

Where technology is global and evolving with lightning speed, regulation is local. Policy, deployment and enforcement mechanisms and instruments have not always shown to be able to adapt, react and govern new developments. With the introduction and global use of the IoT the related challenges will increase.

However, policy making and enforcing such policies, whether being legislation, regulations or otherwise, has valid and very relevant and important reasons and purposes, such as creating, influencing and setting a balanced, predictable trustworthy, fair, reasonable, transparent and open yet where necessary protective framework in order for the society and economy to operate in a trusted and civilised way and be monitored and fostered.

As per technological change, globalisation, worldwide competition and demographical challenges in most regions in the world, operating in a durable hyperconnected economy and society and boosting innovation and productivity are not nice to haves anymore. These are a necessity to have, in order to stay relevant as an economy and society but also to avoid social disruption.

In the Digital Single Market strategy the European Commission basically recognises the importance of these elements within the digital economy.

Within scope of the section on 'Maximising the growth potential of the Digital Economy' of the Digital Single Market strategy, the European Commission has proposed several initiatives to investigate, influence and in some cases propose new or updated standardisation, self-regulation or other policy mechanisms, which will offer new regulatory frameworks. It is part of any society, economy, market or ecosystem, including the IoT ecosystem.

IoT implies a high volume of relationships between many hyperconnected actors – whether human, organisations, algorithms or machines –, and those relationships will need to be arranged and catered for.

These actors within the IoT ecosystem and related digital markets need predictability and legal certainty on the numerous relationships as well as related issues in order to enter the market as vendor or buyer, invest in or procure new products, services and embrace new business models, irrespective of being a private or public organisation or community, or being a governmental body, large corporation, SME or consumer.

In order to make IoT and related hyperconnected ecosystems work, create space to innovate, modernise the society, build global connectivity, nurture internet openness, create trust, jobs and skills in the digital economy and society, and continue working on and safeguard an acceptable level of social prosperity that is durable. The two now colliding worlds of digital technology on the one hand and regulations and compliance on the other will need to be connected and hyperconnected as well.

From an IoT ecosystem and hyperconnected point of view, this Chapter will investigate, point out, explain and structure several of the main challenges, considerations and perspectives in the domain where these worlds meet, collide and will need to get used and adapt to each other in the best way possible.

This Chapter does not exhaustively identify or describe any and all challenges, considerations and perspectives in the domain, and does also not intent to be limiting those challenges, considerations and perspectives mentioned hereunder.

7.2 Unambiguous Definitions

It is fundamental and important to keep the definitions regarding and related to IoT well defined and unambiguous, in order to enable clear communication between multiple stakeholders, to ensure effective recommendations, and to come to a common understanding. Without such basis and common understanding, it is quite impossible to build ecosystems, frameworks, policies and relationships that understand, interact and interoperate with each other in the IoT domain.

As technology and business models develop and new technology and models are developed and adapted, it will also be important to ensure definitions are technology neutral, business model neutral, principle based and consistent with fundamental rights and the fast evolving IoT landscape.

Definition of The IoT by ITU and the IoT European Research Cluster (IERC) is:

> *'The IoT is a dynamic global network infrastructure with self-configuring capabilities based on standard and interoperable communication protocols where physical and virtual "Things" have identities, physical attributes and virtual personalities and use intelligent interfaces and are seamlessly integrated into the information network.'*[1]

The 'Thing' in the IoT can be anything, such as for instance (without limitation) devices, objects, algorithms, people, animals, plants or other Things (hereinafter: *'Things'*). What makes these Things so special is their connectivity via the internet and the ability to act in an orchestrated way, such as Machine to Machine (M2M), Human to Machine (H2M) and Machine to Human (M2H). The combinations are quite indefinite. Taking into consideration that each combination implies as least two 'Things' interacting with each other, there are a lot of legal relationships to address and arrange for.

As the markets and European Commission has currently chosen to use IoT to define this domain, it is good to mention that when one reads or hears about Internet of Everything, Internet of Customers, Internet of Everyone, Internet of Humanity or similar terms, one in essence means the same as the definition of IoT set forth above. However, there is an ongoing debate on whether humans are 'Things'. For the purposes of this book in general and this chapter in specific, it is understood that a human is not a thing but for purposes of setting the scene on legal and other relationships, and in order to easily work with the definition IoT it is within that definition.

In this document, the following terms used shall have the meaning as set forth in the European Commission Cloud Service Level Agreement

---

[1]ITU-T Y.2060, 'Overview of IoT,' June 2012. White paper, 'Smart networked machines and IoT,' Association Instituts Carnot, January 2011.

Standardisation Guidelines[2], ISO/IEC 17788, which guidelines have been initiated, discussed, set and endorsed by the European Commission and currently provide for the most up to date and generally accepted definitions that are to most extent quite relevant and useful in the IoT domain as well.

7.3 Converging Markets

The technologies that make IoT possible are converging existing markets and creating new markets, both physical and virtual markets, private and public markets and both vertical and horizontal markets. From the converging technical markets perspective, smart systems integration, cyber-physical systems, smart networks, data analytics, cloud computing, robotics and artificial intelligence bring together different generic technologies with nano-electronics, wireless networks, low-power computing, adaptive and cognitive systems.[3]

Basically, these can be divided in five main groups:

1. Things
2. Infrastructure
3. Data
4. Services
5. Connectivity and Interoperability

7.3.1 Things

The Things in the IoT are for instance (without limitation) devices, objects, algorithms, people, animals, plants or other Things and are provided with unique identifiers (or sometimes community identifiers) and the ability to transfer data over an infrastructure or network without requiring

---

[2]Cloud Services Level Agreement Standardization Guidelines, European Commission, DG Connect, Cloud Select Industry Group- subgroup on Service Level Agreement (C-SIG-SLA), https://ec.europa.eu/digital-agenda/en/news/cloud-service-level-agreement-standardisation-guidelines

[3]ITU-T Internet Reports, 'IoT', November 2005. Lee, et al. The IoT – Concept and Problem Statement July 2012; F. Mattern and C. Floerkemeier 2010, Mattern, Friedemann, and Christian Floerkemeier, 'From the Internet of Computers to the IoT', in: K. Sachs, I. Petrov, P. Guerrero (red.), *From Active data management to Event-Based Systems and More*, Berlin: Springer 2010, p. 242–259.

human-to-human or human-to-machine interaction. These Things are all about collecting, deriving, using, storing and sharing data via the infrastructure.

7.3.2 Infrastructure

The infrastructure regards transmitting, collecting, storing and/or sharing the data within the ecosystem. It is a collection of hardware, software and other related products and resources that enables the provision of IoT and their services.

7.3.3 Data

The key aspect that keeps IoT moving and alive is data. Data of any form, nature or structure, that can be created, uploaded, inserted in, collected or derived from or within the IoT, including without limitation proprietary and non-proprietary data, confidential and non-confidential data, non-personal and personal data, as well as all other human readable or machine readable data.

Data Life Cycle: the life cycle of processing data commonly includes seven (7) phases:

1. Obtain/collect
2. Create/derive
3. Use
4. Store
5. Share/disclose
6. Archive
7. Destroy/Delete

This data life cycle is also used for personal data, which is then called the personal data life cycle.

It should be noted that data does not only arise out of the first two phases, but data is created and processed in each and any phase. For example, when deleting data, other data describing the act of deletion may arise.

7.3.4 Services

One or more capabilities offered invoked using a defined interface. There is a seemingly endless amount of services offered within IoT in a countless amount of categories, as well as virtual as physical.

The services are extended into fields such as education, intelligent buildings, supply chain, health care, everyday life, disaster management, safety and transport to provide people with better services.

7.3.5 Connectivity and Interoperability

As above-mentioned IoT can be built by using any number of technologies and a particular technology stack should not be assumed. Essential hallmarks of IoT are connectivity and interoperability for which technology neutrality is required.

For example, many products and services are connected with REST interfaces or APIs to exchange data and interoperate with other products, services and Things.

7.4 Relationships and Markets

As the domain of IoT is vast, one needs to identify in which market it is operating and which relationship between which Things it would want to make possible and arrange for. This, as the characteristics of each market and each relationship may have legal consequences and may need specific frameworks and assurances.

The combinations of relationships are endless, as there are quite a few Things, and each combination is possible. For instance business to business, business to consumer, consumer to consumer, government to resident, resident to government, commuters to parking services, and so on.

Whether such relationships make sense, depends on the circumstances and purposes of the relationship and within what market. A whole new range of relationships will arise and on top of that can have multiple purposes.

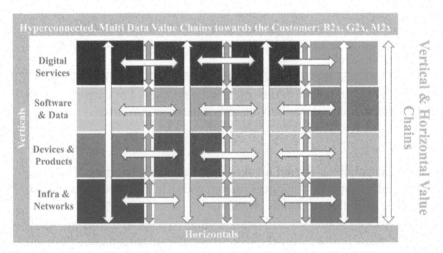

Figure 7.1   Hyperconnected, vertical and horizontal value chain.

For example the development of smart cities and communities is a vertical that has many verticals and cross-verticals combined, for instance without limitation government to business, government to visitors, residents to government, and commuters to business.

As another example, in the business to consumer market one already sees IoT in operation, for example wearables, smart meters, connected cars, smart mirrors and smart fridges. The end-user of IoT in this market is a consumer who uses the IoT in its daily life. It is well-known that a consumer has specific rights to protect its interest, including without limitation personal data protection and product liability.

7.5 What Are the Main Challenges

The Internet, cloud computing, data analytics and other advances in IoT have spawned a global digital economy and the continuing evolution of connected Things has added a new and growing dynamic. While IoT is increasing, it is still in its nascent stages and the related technologies, business models and polices will undoubtedly evolve over a number of years.

There are a number of challenges to facilitate sustainable growth of IoT by adding clarity to the challenges between the converging markets and stakeholders. Several of these main challenges are set below. Again, please note that these are non-exhaustive, and some will be adequately addressed or solved in the near future, where new challenges will surely arise with the emergence of improved or new technologies and combinations thereof.

7.5.1 Common Understanding

It sounds so logical and obvious: we need a common understanding of matters, challenges and solutions. But in fact and real life, it is quite a challenge and in this chapter identified as one of the main challenges.

Common understanding is a result from having found common ground, or a result from having established certain ground as deemed to be the common ground for the matter at hand. When addressing the matters at hand regarding and related to IoT, the same applies. Common ground starts with the basics: clear and unambiguous definitions. Some IoT definitions have been addressed a few paragraphs above. Once the definitions are clear, the next steps are principles and (legal) frameworks that stakeholders recognize and are able to identify themselves with.

Without such definitions, principles and frameworks as a basis for common understanding, it will be impossible to build ecosystems, frameworks, policies and relations that understand, interact and interoperate with each other. The recommendation here is that in case one finds out that a dialogue, discussion, debate or negotiations seem difficult to be resolved, it may be good to take a step back and return to the common understanding, before re-entering into such discussions to seek common ground on the pending matters.

7.5.2 Trust

Trust is always one of the challenges with new technologies and change. Regarding IoT, customers and users thereof may need time to adapt, learn that the opportunities benefits are, and how to mitigate or cope with the risks. Depending on the specific IoT, vertical it is used in, deployment thereof and impact it may have for the customer and users, trust will in some cases be obtained quicker than in other cases.

Integral parts of trust is security, data management, data protection and the way vendors, providers as well as co-users and the related community will act and react on a case to case basis. Another part of building trust is taking care of customers and users with insufficient knowledge. For instance, insufficient knowledge has been established by EuroStat to be the number one reason for businesses not to procure paid cloud services, and the IoT industry should try to avoid that such same barrier arises in the upcoming IoT market.[4]

7.5.3 Security

The technical architecture of the IoT has an impact on security and privacy of the involved stakeholders and data subjects. For example Denial-of-service attacks could be a major threat when it comes to the IoT ecosystem.[5]

Furthermore the security of both the relevant stakeholders in multiple horizontals and verticals is for sure a main challenge as well. The value chains are quite complex in IoT as per its hyperconnectivity and interoperability, which by nature results in customers and users not understanding the possible risks and impact thereof. Even though security is a horizontal itself, it is expressly mentioned as being relevant for other horizontals as well, as IoT

---

[4]Eurostat News Release 9 December 2014.

[5]Denial-of-service-attacks typically involve the overflow of a network device with more requests than it can process, leading to an overload that renders the service unable to answer legitimate requests.

verticals will generally be stacked with several layers, including without limitation infrastructure, networks, products, devices, software, data and services. The value chains of IoT are non-linear towards customers and users; they are multi-angle, cross-vertical, and multi-horizontal. For example, per layer in a vertical IoT ecosystem at least two security horizontal layers may be necessary; one upon entry of data in such layer, and one upon exit thereof.

A high degree of reliability is needed, since business processes are concerned. However there are a lot of similarities when it comes to cloud computing security standards which have already been developed and tested.

Specifying measurable security level objectives in IoT is useful to improve both assurance and transparency. At the same time, it allows for establishing common semantics in order to manage cloud security from two perspectives, namely (i) the security level being offered by a stakeholder and, (ii) the security level requested by a IoT user.[6]

The approach used in this section consists of analysing security controls from well-known frameworks[7] into one or more security objectives, when appropriate. These objectives can be either quantitative or qualitative. This section focuses on the definition of possible security objectives. Eight categories are provided, each with one or more objectives.

The categories represent some important security requirements. However, it should be noted that the list of objectives is not meant to be considered as exhaustive and that the objectives proposed are not meant to be considered as applicable in all individual cases. The applicability of particular objectives depends on the type of products and services offered (in terms of both of functionality and model) and pricing of it (free, paid, premium). It is important to understand that some of the objectives are interdependent: objectives relevant to security may also have relevance in the areas of data management and personal data protection for instance.

- Reliability: reliability is the property of a IoT system to perform its function correctly and without failure, typically over some period of time. The system has to avoid single points of failure and should adjust itself to node failures.

---

[6]Reference is made to Chapter 3 and Chapter 4 of the Cloud Service Level Agreement Standardization Guidelines regarding Security Service Level objectives overview. Furthermore, as an example for the security challenges reference is made to the report 'New security guidance for early adopters of the IoT' of the Cloud Security Alliance which includes an IoT Security Life Cycle.

[7]Relevant security frameworks include in particular ISO/IEC 27001 and ISO/IEC 27002.

- Authentication and Authorization: authentication is the verification of the claimed identity of a Thing. Authorization is the process of verifying that a Thing has permission to access and use a particular data or resource based on the (predefined) ecosystem it wishes to offer to the IoT user. As a principle, retrieved identity and data of a thing must be authenticated.
- Cryptography: cryptography is a discipline which embodies principles, means and methods for the transformation of data in order to hide its information content, prevent its undetected modification and/or prevent its unauthorized use, also known by the term encryption. However stakeholders must be able to implement access control on the data provided in order to cooperate within the IoT ecosystem.
- Security incident management and reporting: an information security incident is a single or a series of unwanted or unexpected information security events that have a significant probability of compromising operations and threatening information security. Information security incident management are the processes for detecting, reporting, assessing, responding to, dealing with, and learning from information security incidents.
- Logging and Monitoring: logging is the recording of data related to the operation and use of a IoT system. Monitoring means determining the status of one or more parameters of a IoT system. Logging and monitoring are ordinarily the responsibility of the relevant stakeholder's.
- Vulnerability Management: a vulnerability is a weakness in an IoT system, security procedures, internal controls, or implementation that could be exploited or triggered by a threat. Management of vulnerabilities means that information about technical vulnerabilities of information systems being used should be obtained in a timely manner, the exposure to such vulnerabilities evaluated and appropriate measures taken to address the associated risk.
- Governance: governance is a framework by which IoT will be directed, controlled and governed.

7.5.4 Personal Data Protection

Data used to be quite static, and used to reside in one place. Digital data that is connected to internet and cloud, and that is hyperconnected through IoT ecosystems, travels. This may be the key catalyst of internet, cloud computing, IoT and data analytics technologies being in some kind of way used and embraced by each and any organisation in the worlds. The fact that data travels is not new, but have come on the agenda in the past years of both the demand side as the vendor side, as well on the agenda of policy makers. As data subjects, data controllers, companies, organisations and countries feel they are losing control over their respective data, and do not always understand or know how the data is processed, it is only natural that some of those are reacting to try to regain control, whether it is personal data, sensitive data or otherwise.

New regulations and directives related to personal data protection, such as the General Data Protection Regulation (GDPR), and security breach notifications, such as the Network and Information Security Directive (NIS Directive) add to those concerns, but may also be part of the keys and mechanisms to resolve these concerns, if implemented in a transparent and understandable way for such data subjects, customers and users.

To understand (personal) data protection it is recommended to go back to the basics, which means that data is not a four letter word. The difference between the definitions of data and personal data should be clear and a common understanding. Reference is made to paragraph 7.3.3 for the definition of data where data is explained and what kind of data there are in the IoT ecosystem.

Personal data means any information relating to an identified or identifiable natural person ('data subject'); an identifiable person is one who can be identified, directly or indirectly, in particular by reference to an identification number or to one or more factors specific to his physical, physiological, mental, economic, cultural or social identity.[8] For the protection of personal data the employment of technical, organisational and legal measures in order to achieve the goals of data security (confidentiality, integrity and availability), transparency, intervenability and portability, as well as compliance with the relevant legal framework is required.

The basis for personal data should be data minimization, where stakeholders are responsible for ensuring that personal data is erased (by the provider and any subcontractors) from wherever they are stored as soon as they are no longer necessary for the specific purposes.

Based on the (new) regulations and directives related to personal data, the principle of purpose specification and limitation requires that personal data must be collected for specified, explicit and legitimate purposes and not further processed in a way incompatible with those purposes. Therefore, the purposes

---

[8]Chapter 2 and Chapter 6 of the Cloud Service Level Agreement Standardization Guidelines regarding personal data protection.

of the processing must be determined, prior to the collection of personal data, by the data controller, who must also inform the data subject thereof. When the data controller decides to process data in IoT, it must be ensured that personal data is not (illegally) processed for further purposes by the relevant stakeholder, or one of the subcontractors.

Only if the data controller informs the data subject (the IoT user) about all relevant issues and being transparent which data will be collected, then the data controller is allowed and to process any (personal) data. The reason therefore is that the data subject is capable of fulfilling its obligation to assess the lawfulness of the processing of personal data. Moreover, the data controller shall make available the information that enables the customer to provide the data subjects with an adequate notice about the processing of their personal data, as required by law. Furthermore, the (new) regulations and directives related to personal data gives the data subject the right of access, rectification, erasure, blocking and objection.

One of the (new) requirements for the data controller as set forth in the GDPR are codes of conduct, standards and certification mechanisms. Such codes of conducts, standards and certification mechanisms for the stakeholders gives more and clear guidelines how the IoT user is protected. The data controller, must accept responsibility for abiding by the applicable data protection legislation.

In order to maintain the above (personal) data protection approach within IoT, the key is to adhere privacy-by-design in advance. In accordance with the laws and regulations regarding data protection and the challenge thereof within IoT, the main principles for privacy-by-design are the following:

- No personal data by default principle: avoid personal data collection or creation by default, except where, when and to the extent required.
- 'As-If' principle: design and engineer IoT ecosystems as-if these will process personal data, now or in a later phase.
- De-Identification by default principle: de-identify, sanitise or delete personal data as soon as there is valid legal basis anymore.[9]
- Data minimalisation by default: only process data where, when and to the extent required, and delete or de-identity other data.
- Encryption by default principle: encrypt personal data by default, and include digital rights and digital rights management thereto.

---

[9]For more information on this topic please refer to 'NIST Special Publication 800–88: Guidelines for Media Sanitization'.

7.5.5 Digital Right Management

As companies transition to IoT, the traditional methods of securing and managing data is challenged by IoT-based connections. Elasticity, multi-tenancy, new physical and logical infrastructures, and abstracted controls require new data security strategies. Managing data and information in the area of IoT can affect all organisations, users and Things. It begins with managing internal data and service migrations and extends to securing information in diffuse, cross-organisation applications and services.

The data management objectives cope with important quantitative and qualitative indicators related with data life cycle management, and can be considered as complementary to existing and applicable security and data protection certifications offered by the IoT stakeholders.

The presented data management objectives are subdivided in four (4) different top-level categories covering all aspects of the identified data lifecycle. Each category is subdivided in one or more objectives that are applicable to that specific category. Not all objectives may be relevant for each service, in particular depending on the type of Things and stakeholders as M2M, M2H or H2M.

7.5.6 Data Ownership and Data Access

Who owns the data? Why am I not able to retrieve my data? From the customer and users perspective, the existing awareness, expertise and transparency of both such customers, users as well as vendor level as well as policy makers and authority level is generally not sufficient to provide actors in the data value chain with trust, predictability and legal certainty each needs to be able to assess, make informed decision and have reasonable access and use of IoT and related services. The same goes for the existing legal frameworks and current contractual practices, although this obviously differs per product, deployment model and service model as well per vendor and the (envisioned or actual) use of the customer and users thereof. Contractual practices, including the arrangements in or related to IoT also create obstacles to data use, access, and in quite a few places create data lock-in effects as well.

One of the challenges with data ownership is that the concept of 'owning' digital data in the traditional sense of the word ownership in most cases in an oxymoron and leads to discussion and conflicts. Data ownership is generally not addressed in the IoT domain, because data ownership is a difficult domain, also as it is not defined. Vendors may have a totally other opinion or perception about data ownership than its customers and users, whether being SMEs or not, and the laws and regulations that have deemed to be governing ownership are either outdated or are quite difficult to apply, interpret, use and enforce in the digital world.

It becomes even more problematic when some vendors have a traditional mind-set that owning assets, including data, is a goal in itself. From another perspective, ownership of digital data in general is basically not possible. The current framework of copyright regulations is not particularly designed for digital assets including data, while the redesign thereof in the early 90s regarding software (Directive 91/250/EEC) has not proved to be a transparent framework that resolves discussions and disputes on ownership as well. The Database Directive (96/9/EC) from 1996 also has lost its effectiveness, as a major requirement for protection thereunder is having done a substantial investment to build the relevant database, where such databases nowadays can be built and used for a fraction of the cost. The threshold to be eligible for protection thereunder is not met anymore, and lowing the threshold would even increase and not resolve the discussion on data ownership either. The upcoming Trade Secret Directive (COM/2013/0813) that is proposed, may resolve a minor part of the data ownership discussion, but in such case the protected data thereunder needs to remain secret and not generally known or readily accessible to third parties. In hyperconnected ecosystems where data travels and data can change from legal characteristics and purpose of travelling and being processed at any time, this will be quite challenging. Owning data is just very difficult, as one would like, or need to, share such data, have it processed and transferred. On the other hand, domain names and related domain name rights have been designed by law not to use the concept of ownership; it uses the concept of holdership of a domain name, which has proven to work quite well. Based on research done and ongoing research by Arthur's Legal, introducing and using the term 'data control' is the preferred way to move forward in the IoT and related digital and hyperconnected domains. This also to reflect on the challenges set above and to address the confusion and distrust that the term 'data ownership' leads to. Data control better reflects the rights a person or organisation may have, whether personal data or non-personal data, and the rights can grant others. It also reflects that digital data can and in most cases will be shared and processed. Data control will be one of the most relevant and essential components to boost trusted hyperconnectivity and the digital economy and society, as it is all about data.

The European Commission has data ownership and data-access on its agenda, and has started the dialogue about how to be able to address this domain of use rights and digital rights management. As Commissioner Oettinger put it on these topics: 'We need a single rule book for the Internet of Things in Europe. Capable to properly address new challenges raised by the technology. This includes data protection, safety and liability rules, including the emerging issues of data ownership, rules on access and re-use of non-personal data in an industrial context, just to mention a few.'[10]

7.5.7 Free Flow of Data

On the Free flow of data, it can be established that restrictions on the free movement of data within the European Union and unjustified restrictions on the location of data for storage or processing purposes are generally not addressed in generic IoT products and services. This is understood as most restrictions are only applicable to certain industries, markets or use. It is however a main challenge as hyperconnected ecosystems are borderless and the data therein should be able to flow freely and unrestricted, at least within the European Union.

Quite a few member states have implemented sector-specific rules and regulations that differ per member state, thus hampering the digital single market and European manufacturers, services providers and other vendors to benefit from being able to market its respective products, services and data to other member states.

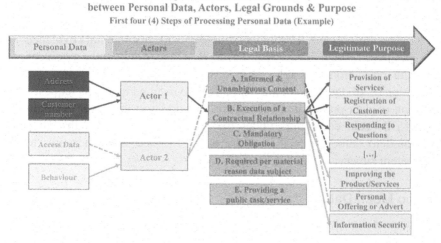

Figure 7.2 Example of data relation flows.

---

[10]http://ec.europa.eu/commission/2014-2019/oettinger/announcements/keynote-speech-closing-plenary-session-net-futures-2016-brussels_en

7.5.8 Accountability and Liability

As per the convergence of technologies, markets and stakeholders, and as these technologies, its manufacturers and vendors are diverse, the question who is accountable and liable for what will be even more difficult to answer and proof than in the physical society.

For example, a manufacturer of certain objects has to accept and address its respective and proportionate responsibility in the IoT ecosystem its objects are deployed. IoT will bring more responsibility for each stakeholder in the market, and each of such stakeholders will have to think and arrange for those effects in a transparent, diligent and ethical manner.

Another example is a security breach in an IoT ecosystem as per insecure coding of software somewhere in the multi-angled value chain. As long as related software companies cannot be held accountable, a solid and stable digital economy and society will be difficult to create.

Merely contractually re-allocating risks and damages to the customer and its users will not contribute to the creation of the Digital Single Market in general, and uptake the IoT ecosystems in particular.

In the field of data protection, accountability often takes a broad meaning and describes the ability of parties to demonstrate that they took appropriate steps to ensure that data protection principles have been implemented.

In this context, accountability is particularly important in order to investigate personal data breaches; to this end, the relevant stakeholders should provide reliable monitoring and logging mechanisms.

Moreover, the relevant stakeholders should provide documentary evidence of appropriate and effective measures that are designed to deliver the outcomes of the data protection principles (e.g. procedures designed to ensure the identification of all data processing operations, to respond to access requests, designation of data protection officers, etc.). In addition, IoT users that are deemed to be data controllers under the GDPR should ensure that they are prepared to demonstrate the setting up of the necessary measures to the competent supervisory authority upon request.

7.5.9 Too Much Data?

The billions of sensors and other objects and Things will generate so much data, most of which is expected to be unstructured and not necessarily useful yet making identifying relevant data more difficult. Commonly available data analytics technologies cannot yet cope with the amount thereof in a comprehensive, useful way. As data analytics is one of the pillars to make

IoT interesting, feasible and worthwhile, this can be seen as one of the main technological challenges. Data architecture will therefore be quite important to address. This, for instance also to comply to regulations and standardisation including without limitation regarding personal data protection, security breach notification and the like.

7.5.10 Regulation and Standardisation

As argued in the paragraph 7.1, technology is global and regulation is local, and new technology goes to market much quicker than regulations.

Policy makers are investigating and deploying other policy mechanisms such as industry guidelines, best practices, codes of conducts, international standardisation, community self-regulatory initiatives and the like, to find the right hybrid combination to be able to adapt, react and govern such technological and related developments.

Getting the right mix of policies in the market, in time yet in a durable and facilitating way, is quite a challenge nowadays.

7.6 Multi-Angle Stakeholders IoT Ecosystem

IoT can be built using any number of technologies, used by different stakeholders and for all kind of markets, whereby all kind of goals should be formalized and covered by ethics, accountability, standardisation, legislation agreements and insurance. Essential to reach this goal of IoT is that these are based on technology neutral wording as a necessary foundation.

7.6.1 Technology and People

The most important elements of the multi-angle stakeholders IoT ecosystem are the technology and the use thereof by the people. The technology of IoT should be neutral and monitored from time to time to be up to date and based on the state-of-the-art technology. Monitoring of the technology will be based on the principles of security, personal data, digital right management, usability, portability and accountability. If technology and for instance IoT will be a success the people have to accept such new connected technology. People play one of the key roles in the Ecosystem especially when it comes to implementation, acceptance and trust of IoT. The human factor is a big challenge of IoT. Both technology and people have influence on the other stakeholders of IoT as ethics, accountability, regulation, standardization, legislation and risk allocation, which will be further discussed in the paragraphs below.

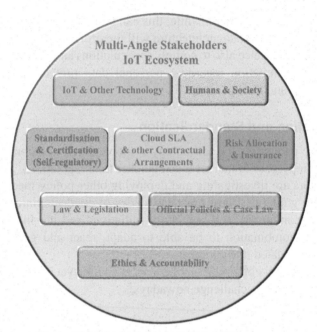

Figure 7.3   Multi-angel stakeholders IoT ecosystem.

7.6.2 Ethics and Accountability

In every converging market, each stakeholder has to deal with ethic and accountability regarding to IoT. For example, a manufacturer of certain Things has to address the challenges of ethic with the connected Things, and whereby such manufacturer has to accept its responsibility for all subjects of IoT. In case of safety of a product, it is not possible to cover all product liabilities, but IoT will bring more responsibility for each stakeholder in the market, and have to think and discuss those effects in a transparent manner prior to the other goals of IoT.

7.6.3 Regulation and Standardisation

The internet is a global communication channel and it is built on standards that are respected worldwide, which is also the basis of IoT. However, the government and compliancy of Things is covered by the current legislations, which will not fit completely for purposes of IoT. There are regional, national and local laws have to govern the use of Things and all other aspects of

IoT for all kind of converging markets, stakeholders and markets. The goal is that legislation should fit for IoT based on technology neutral legislation, because everyone benefits from globally common understanding vocabulary and legislation.

Standards and guidelines for IoT should specify the concepts and definitions necessary for the converging markets, stakeholders and markets to describe the Things, infrastructure, data and services life cycles. There are already standards and guidelines used and produced by organisations such as ENISA, NIST or ISO/IEC. For example, in the field of security, relevant work is using the approach to analyse and refine an individual control into one or more security objectives, which are then associated with metrics and measurements that can be either quantitative or qualitative. Before introducing a particular concept into a standard or guideline for IoT, one should seek proof to ensure the concept is viable from both technical and business perspectives. With standardization of guidelines in the relevant markets of IoT, it will create world-wide applicability, technology and business model neutral, unambiguous definitions and create conformance through a global, common understanding.

7.6.4 Contractual Relationships

The agreement between the stakeholders can refer to the clearly defined information in the legislation, standards and guidelines, but the agreement itself must meet local legal requirements and those must be left to the discretion of qualified attorneys.

7.6.5 Risk Allocation

All the other risks, liabilities and other elements which could not be defined and arranged by legislation, standardisation and agreement should in a best case scenario be covered by insurances. If insurance of IoT is possible than it realizes that IoT is a mature market and the IoT ecosystem is complete.

7.7 Conclusion and Recommendations

New technologies lead to change. Change is a catalyst that can be feared, but can also be embraced and used to optimise the current status quo of society and economy, and sometimes even leapfrog technologies that have already been improved. Especially the hyperconnected aspect of IoT technologies will have quite some impact on the society and economy, and may raise certain ethical or legal discussions on new and existing topics.

As the IoT technologies, developments, combinations and deployments of IoT verticals and horizontals continue to evolve, the opportunities and challenges will evolve as well, including the legal and compliance consequences, challenges and opportunities, including privacy, security and other compliance by design and related automation thereof. Addressing and resolving these challenges are one the most important value creating and success factors of IoT. On privacy-by-design, the Article 29 Working Party worded it as follows in the Opinion 8/2014: 'Organisations which place privacy and data protection at the forefront of product development will be well placed to ensure that their goods and services respect the principles of privacy-by-design and are equipped with the privacy friendly defaults expected by EU citizens'.

As any relatively new market, also the IoT supply side and IoT demand side will need to find, understand an trust each other the coming period, and for that a principle-based ecosystem of IoT policy frameworks may facilitate of uptake of that market. As its hyperconnected, agile and hybrid nature, such policy framework ecosystem will need to be hyperconnected, agile and hybrid as well in order to have the positive impact it seeks. Principle-based mechanisms with a solid common ground of globally recognised definitions and principles will facilitate such agile framework ecosystem. Each policy framework will need to be hybrid, with all the tools and mechanisms available and newly developed, including for instance self-regulation, community frameworks, standardisation, where relevant current regulation and where necessary regulation, preferably Pan-European because of the borderless nature of technology. As per the extreme variety of actors, objects, markets, capabilities, Things and relationships, one single IoT framework seems difficult, hence the conclusion that an interoperable and durable ecosystem of IoT framework may be the way to facilitate and support the market and all related stakeholders in an efficient way. Such ecosystem will need to be based on open and transparent dialogues with a large variety of groups and stakeholders from a 3D multi-angle, both internally at the European Commission, as well as externally in the European Union and beyond.

With such hyperconnected multi-disciplinary brainpower and related combinatorics innovation, trust, usability and market update will have the best chances to succeed, and may result in multiplicity: a symbiotic combination of diverse groups of people working together with diverse groups of machines to make decisions and solve complex problems. As Commissioner Oettinger put it on a human-centred IoT: 'The aim is to empower citizens rather than machines and corporations, thanks to high data protection and security standards.'[11]

With that, the IoT combined with the Internet of Humanity leads to the Internet of Human Prosperity.

---

[11] http://ec.europa.eu/commission/2014-2019/oettinger/announcements/keynote-speech-closing-plenary-session-net-futures-2016-brussels_en

8

IoT Standards – State-of-the-Art Analysis

**Emmanuel Darmois[1], Omar Elloumi[2], Patrick Guillemin[3]
and Philippe Moretto[4]**

[1]CommLedge, France
[2]Nokia, France
[3]ETSI, France
[4]UNIFY-IoT, ESPRESSO smart cities and Sat4m2m, Germany

8.1 Introduction

The Internet of Things (IoT), as an emerging technology, has the potential to boost innovation in many industrial sectors, as well as to help address many societal challenges including climate change, resource and energy efficiency and ageing.

However, this potential will only materialize if IoT develops as an open platform cutting through the silos, supporting a variety of applications and generating open and sustainable ecosystems. As for any new technology as it begins to emerge, there are many proprietary or semi-closed solutions together a number of existing – and somehow competing – standards, thus re-enforcing the perception that the IoT landscape is fragmented, in particular for standards.

The objective of this chapter is to present the main initiatives that contribute to the analysis of the current status and dynamics of the IoT standardization and to outline their early findings.

The main contributors of this analysis of the IoT standards landscape are the AIOTI WG03, the ETSI Specialist Task Force (STF) 505 and the UNIFY-IoT Coordination and Support Action (CSA).

8.2 Analysing the IoT Standards Landscape

This section introduces three initiatives that target the analysis and under-standing of the IoT standards landscape. They are briefly described below and their respective contributions – reference models, identification of Standards Developing Organizations (SDOs), collection of existing standards, and identification of gaps – are further referred to in the rest of the chapter.

8.2.1 AIOTI WG03

Within the Alliance for IoT Innovation (AIOTI, www.aioti.eu), AIOTI WG03 (IoT standardization) is a focus point of European engagement and steering in the standardization process. In collaboration with other AIOTI Working Groups and STFs, WG03 is:

- Maintaining a view on the landscape of IoT standards-relevant activities being driven by SDOs, Consortia, Alliances and OSS projects.
- Providing a forum for analysis, discussion and alignment of strategic, cross-domain, technical themes and shared concerns across landscape activities.
- Developing recommendations and guidelines addressing those concerns.

AIOTI WG03 is engaging the IoT community in disseminating and promoting its results and steering emerging standards. In particular, WG03 is expected to play an important role in conjunction with:

- The upcoming IoT Large Scale Pilots (LSP) (http://ec.europa.eu/research/ participants/portal/desktop/en/opportunities/h2020/topics/2223-iot-01-2 016.html) and the IoT LSP large CSAs.
- Other initiatives and cross alliances that have been prepared and will be launched after the AIOTI General Assembly's approval on 30 May 2016.

AIOTI WG03 has undertaken parallel tasks on 1) IoT Landscaping; 2) IoT High-Level Architecture; 3) IoT Semantic Interoperability; 4) IoT Privacy. A first version of the results of this work has been published in November 2015 together with the other AIOTI WGs reports (http://bit.ly/1GtzJ5I). An updated version of the AIOTI WG03 IoT landscaping report (version 2.6) has been published at the end of May 2016.

8.2.2 ETSI STF 505

STF505 (https://portal.etsi.org/stf.aspx?tbid=595) is a group of experts, funded by the European Commission under the rolling plan on ICT stan-dardization in collaboration with the European Multi-Stakeholder Platform (MSP) and supported by ETSI, commissioned to provide on the one hand an in-depth analysis of the IoT standardization landscape (in particular

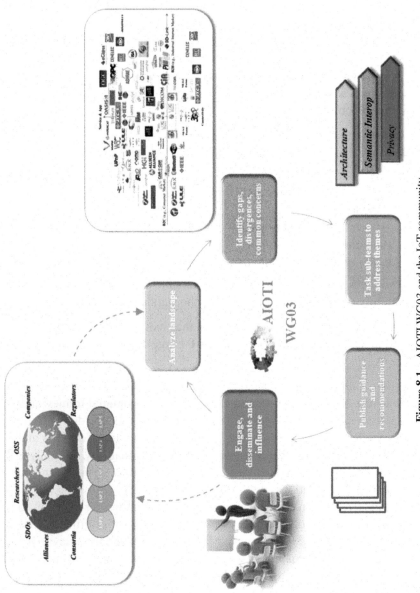

Figure 8.1 AIOTI WG03 and the IoT community.

in conjunction with the AIOTI), and on the other hand, an identification of the IoT standardization gaps.

The study considers "vertical" functionalities (standards and protocols) in specific application domains, i.e., a single vertical industry, such as home automation, smart mobility and wearable medical devices, etc. and "horizontal" functionalities that are not specific to any particular domain but aim to provide common standards, protocols and solutions applicable to as many vertical industries as possible.

The essential objectives are:

- To analyse the status of current IoT standardization.
- To leverage liaisons between SDOs, SSOs, and industrial alliances, which allows:
  - To assess the industry and vertical market fragmentation vis-à-vis standardization
  - To point towards actions that can increase the effectiveness of IoT standardization, to improve interoperability, and to allow for the building of IoT ecosystems.
- To foster dissemination work for the sustainable development of a global community of stakeholders involved in the standardization of IoT.

This STF is developing a set of deliverables that will include recommendations aimed at supporting material for the IoT Large-Scale Pilots (LSPs), in particular:

1. A technical report (TR) on standards landscape for IoT (who does what, what are the next milestones) and identification of potential frameworks for interoperability (e.g. oneM2M). The methodology for drafting the TR on "IoT standards landscaping" is to collect and analyse SDOs and industry standards, to evaluate their stability and maturity, to analyse complementarities/antagonism with open source development, and to provide recommendations.
2. A technical report on identification of gaps and proposals on how to address them in standardization. The methodology for drafting the TR on "IoT European LSP gap analysis" is to systematically analyse the SDOs standards and roadmaps with the support of a survey, to map the LSPs use cases and lifecycle on the related standards, and to identify gaps in standard support;
3. A thematic workshop on Smart Home covering different LSP application domains such as Smart Living, eHealth, Wearables and Smart Cities;

4. An extended delivery workshop centred on the presentation of the STF results and their application by the winning LSP proposals.

8.2.3 UNIFY-IoT CSA

The objectives of UNIFY-IoT (www.unify-iot.eu) are to stimulate the collaboration between IoT projects, between the potential IoT platforms (http://iot-epi.eu/) and to support them in sustaining the IoT ecosystems developed by focusing on complementary actions. The overall goal is to foster and stimulate acceptance of IoT technology as well as the means to understand and overcome obstacles for deployment and value creation.

The UNIFY-IoT CSA is aiming to be a "working partner" of AIOTI and the IoT European Research Cluster (IERC) by coordinating and supporting the activities on innovation ecosystems, IoT standardization, policy issues, research and innovation, and goes beyond the classical workshops and conferences.

The cooperation framework consists of six dedicated Tasks Forces: Innovation, Platforms interoperability, IoT accelerators, IoT business models, Educational platforms and International cooperation.

UNIFY-IoT will also establish links with the selected H2020 IoT-01-2016 LSPs and IoT-02-2016 CSAs when they will start to operate at the beginning of 2017.

8.3 A Framework to Analyse IoT Standardization

In order to ensure a sound and safe methodology to capture the standards landscape and to identify the standardization gaps, some elements have been used as a framework for the analysis of the IoT landscape and are described in this section. Some of them come from the AIOTI WG03 work or from the STF 505 work and some are a refinement of the AIOTI WG03 work made by the STF 505.

8.3.1 Horizontal and Vertical Domains

The AIOTI WG03 has defined a way to graphically represent the "IoT SDOs and Alliances landscape (vertical and horizontal domains)" by highlighting the main activity of SDOs and Alliances with respect to two dimensions:

- The IoT application domains represented as "verticals"
- The IoT telecommunication infrastructure domain represented as "horizontal/telecommunication".

8.3.1.1 Vertical Domains

The vertical domains are those that are addressed by the IoT LSPs as described below:

- *Smart cities*. The modern cities need to evolve and become structured and interconnected places where all components (energy, mobility, buildings, water management, lighting, waste management, environment, etc.) are working together for the benefit of humans. By using IoT technologies, the cities are expected to achieve this transition while maintaining security and privacy, reducing negative environmental impact and doing it in a reliable, future proof and scalable manner.

- *Smart living environments for ageing well* (e.g. smart home). It is expected that the IoT will support the continuously growing population of elderly people in living longer, staying active, non-dependent and out of institutional care settings, together with reducing the costs for care systems and providing a better quality of life. This should be achieved in particular with IoT for smart home and home automation supporting technologies.

- *Smart farming and food security*. The application of IoT technologies to the overall farming value chain will enable to produce more with less resources and negative environmental impacts and will improve at the same time food safety. Technologies such as data gathering, processing and analytics as well as orchestrated automation technologies supported by IoT are expected to achieve this.

- *Wearables*. The integration of intelligent systems to bring new functionalities into clothes, fabrics, patches, aids, watches and other body-mounted devices will provide new opportunities and applications. Basic technologies such as nano-electronics, organic electronics, sensing, actuating, localization, communication, etc. will be offered to the end-user, with an associated range of problems such as acceptability, ease of use, privacy, security or dependability.

- *Smart mobility (smart transport systems/smart vehicles/connected cars)*. The IoT applied to the mobility domain may create the potential for major innovations across a wide variety of market sectors, with mobility applications such as self-driving and connected vehicles, multi-modal transport systems and "intelligent" transportation infrastructure from roads or sea ports to parking garages.

- *Smart environment (smart water management)*. IoT will be a key building block to solutions for vertical applications such as environmental monitoring and control that will use sensors to assist in environmental protection by monitoring air, water quality, atmospheric or soil conditions and noise pollution.

- ***Smart manufacturing***. In support of the European manufacturing industry and of the Factories of the Future, all forms of competitive industries will have to massively incorporate more intelligence, that will rely in particular on IoT through advanced connected objects providing sensing, measurement, control, power management and communication, both wired and wireless.

The vertical domains are further used for the classification of the identified standards and standardization gaps.

Note: After the final definition of STF 505, AIOTI has included two new Working Groups on "Smart Buildings and Architecture" and "Smart Energy" that should be included in the future extensions of the scope of the IoT Gap Analysis.

8.3.1.2 IoT SDOs and Alliance Landscape

The "IoT SDOs and Alliances landscape (vertical and horizontal domains)" representation generated by the AIOTI WG03 is illustrated in Figure 8.2 below, in its release 2.6.

This representation captures the SDOs and Alliances that are active in the horizontal and vertical domains.

It should be noted that this representation includes not only SDOs (that develop standards for the IoT) but also alliances that very often serve different purposes such as marketing, promotion, solution profiling, etc.

8.3.2 Knowledge Areas

The AIOTI WG03 has defined Knowledge Areas (KAs) that are further used for the classification of Standards in the subsequent sections of this chapter. The definition of the AIOTI WG03 KAs is reminded below (with minor adaptations made by the STF 505):

Communication and Connectivity

This KA covers mainly specification of communication protocols at all layers, e.g., PHY, MAC, NWK, Transport, Service, and Application layers. It includes the management associated with the KA.

Integration/Interoperability

This KA covers mainly specification of common IoT features required to provide integration (assembly of sub-systems) and interoperability (interoperation of heterogeneous sub-systems).

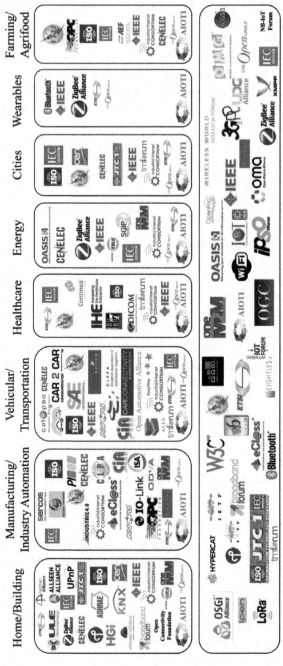

Figure 8.2   IoT SDOs and alliance landscape by domains.

Applications

This KA covers the support of the applications lifecycle. This includes development tools, application models, deployment, monitoring and management of the applications.

Note: The application level protocols, APIs, data models, ontologies, etc. are part of the "Communication and Connectivity" KA.

Infrastructure

This KA covers the design, deployment, and management of computational platforms and infrastructures (e.g. network elements, servers, etc.) that support IoT-based usage scenarios.

IoT Architecture

It covers the specification of complete IoT systems, with a focus on architecture descriptions.

Devices and sensor technology

This KA covers mainly device and sensor lifecycle management.

Note: The communication protocols between devices and other elements are covered in the "Communication/Connectivity" KA.

Security and Privacy

This KA covers all security and privacy topics.

8.3.3 High Level Architecture (HLA)

8.3.3.1 The AIOTI HLA

The AIOTI WG03 has developed the AIOTI High Level Architecture for IoT (AIOTI HLA), a standard framework or architecture for IoT that should be applicable to IoT LSPs. The HLA is meant to be the basis for further discussion with the LSP WGs in order to promote architectural convergence among the WGs.

The AIOTI HLA is similar or can be mapped to other frameworks such as those developed by ITU-T, oneM2M or IIC. An example of such a mapping is provided in the next sub-section.

The purpose of AIOTI HLA (and of the other frameworks) is in particular to support interoperability in complex IoT systems and to provide means of identifying and defining interworking standards with reduced complexity. This framework also supports the characterization of standards gaps and is used by the STF 505 to this extent. A functional model of the AIOTI HLA is depicted in the Figure 8.3 below.

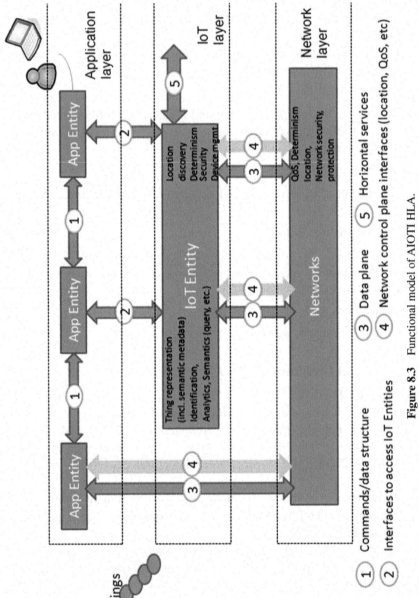

Figure 8.3   Functional model of AIOTI HLA.

The role of the interfaces is to:

- Defines the structure of the data exchanged between app entities (the connectivity for exchanged data on this interface is provided by the underlying networks). Typical examples of the data exchanged across this interface are: authentication and authorization, commands, measurements, etc.
- Enables access to services exposed by an IoT entity to e.g. register/ subscribe for notifications, expose/consume data, etc.
- Enables the sending/receiving of data across the networks to other entities.
- Enables the requesting of network control plane services such as: device triggering (similar to "wake on LAN" in IEEE 802), location (including subscriptions) of a device, QoS bearers, deterministic delivery for a flow, etc.
- Enables the exposing/requesting services to/from other IoT Entities. Examples of the usage of this interface are to allow a gateway to upload data to a cloud server, retrieve software image of a gateway or a device, etc.

8.3.3.2 Mapping of the HLA: The Example of oneM2M

This fragmented nature of the M2M and IoT market has led to the creation of oneM2M, an alliance of standards organisations looking to develop a single horizontal platform for the exchange and sharing of data among all applications.

oneM2M is creating a distributed software layer, like an operating system, which is facilitating that unification by providing a framework for inter-working with different technologies, e.g. OMA LWM2M, OCF (previously OIC) and AllSeen (providing standards for proximity networks). oneM2M enables interoperability across IoT applications regardless of the underlying technology used.

oneM2M defines a Common Services Layer which is a software layer that sits between the network and applications, be it in the wide area network domain or in the filed domain (where devices and gateways are generally deployed). The functions in the Common Services Layer include: device management, data collection, protocol conversion and interworking, group management, security, etc. Those functions are exposed to applications (in the cloud, gateway or devices) via Restful APIs.

The functional oneM2M architecture is depicted in Figure 8.4. The reference points defined by oneM2M are:

- **Mca**: is a Restful API to expose functions in the common service layer to applications running in the devices, gateways or in the cloud.
- **Mcc**: is the reference point that allows, among other features, to register a device or gateway, perform discovery, and exchange IoT data on behalf of the applications. Mcc is also used to perform device management that allows managing the lifecycle of the devices. Device management functions include: software and firmware upgrade, configuration management and performance management.
- **Mcn**: is the reference point that allows to access network services, on behalf of IoT applications, such as device triggering (for sleeping devices), network location or request quality of service connections for IoT services that need specific guarantees.

The Figure 8.5 provides a mapping between oneM2M and the AIOTI HLA functional model.

oneM2M has specified all interfaces depicted in Figure 8.4 to a level that allows for interoperability. Three protocols binding have been specified, so far, for Mcc and Mca reference points: CoAP, MQTT and HTTP. As regards the Mcn reference point, normative references have been made to interfaces specified by 3GPP and 3GPP2 in particular.

However, oneM2M does not specify vertical specific data formats for exchange between App Entities according to AIOTI HLA interface 1. This can however be achieved by interworking with other technologies such as ZigBee, AllSeen, etc.

8.3.3.3 The STF 505 Enterprise IoT Framework

The STF 505 has defined an Enterprise IoT Framework in order to put a global structure on the framework used for the analysis of the SDO landscape. Such a framework has to deal not just with the technology, but also with other relevant area to be taken into consideration such as include the stakeholder views, the regulatory aspects (e.g. for a city): all these make up an enterprise view.

The AIOTI WG03 reports point out that part of the complexity of IoT comes from its intention to support a number of different applications covering a wide array of disciplines that are not all part of the ICT domain: taking an overview of all these elements can be overwhelming without a structural view. The STF 505 approach is to view the IoT framework as an Enterprise Architecture (in line with the TOGAF model for Enterprise Architecture).

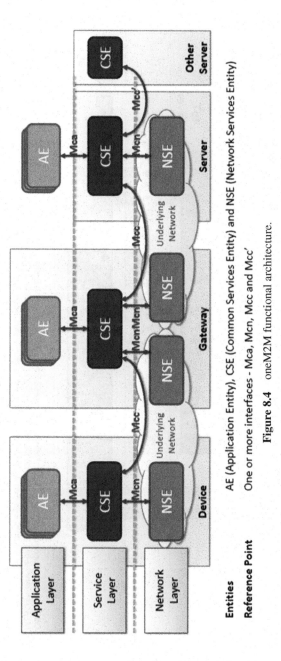

Figure 8.4 oneM2M functional architecture.

Entities   AE (Application Entity), CSE (Common Services Entity) and NSE (Network Services Entity)

Reference Point   One or more interfaces - Mca, Mcn, Mcc and Mcc'

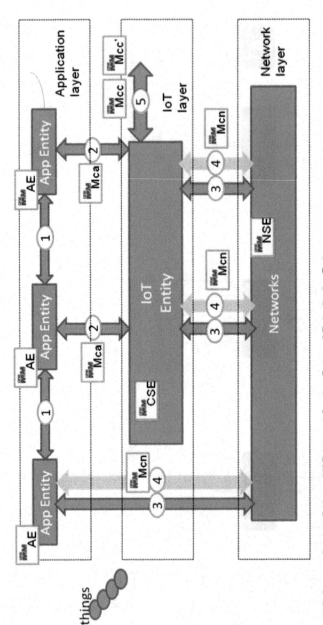

CSE: Common Services Entity - **NSE**: Network Services Entity - **AE**: Application Entity

Mcn: reference point between a CSE and the Network Services Entity (NSE), enable a CSE to use network services such as location and QoS

Mcc/Mcc': reference point between a CSE and a CSE. It allows registration, security, data exchange, subscribe/notify, etc.

Mca: API to Application Entities that expose functions of the CSE.

oneM2M CSE functions include: device management, registration, discovery, group management, data management and repository, etc.

Figure 8.5 Mapping oneM2M to AIOTI HLA.

The main elements of this framework shown in Figure 8.6 are the following:

- An architecture reference model which consists of an IoT architecture integrating all components that make up an IoT system;
- An IoT domain which holds the view of what makes up an IoT system;
- A standards information database which is the main object of study of the IoT standards landscaping, aiming to hold all relevant standards that can be used;
- A reference library which holds any re-useable information that can be used across the IoT LSP pilots;
- A governance repository that houses all policies, regulations that applies to any LSP.

The Standards Information Database

Standards are published documents that establish specifications and procedures designed to ensure the reliability of the materials, products, methods, and/or services people use every day. The Standards Information Base (SIB) collects the standards with which new architectures must comply, which may include industry standards.

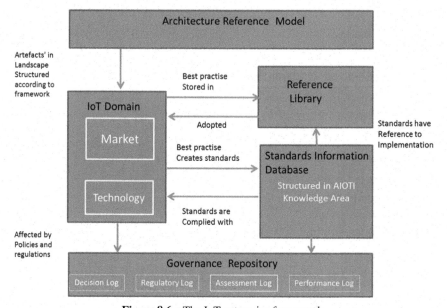

Figure 8.6 The IoT enterprise framework.

AIOTI have come up with a set of standards that are relevant to the LSPs. Some of the standards are common across the LSPs, while some are specific for a LSP. In order to better represent the standards landscape across the various technology areas, AIOTI have come up with the concept of "knowledge area" that has been used a classification scheme for the standards information base.

Standards and Knowledge Areas

Considering the KAs defined above, it may be appropriate to clarify the kind of standards that can be expected in a given KA. Some examples are given below. STF 505 has used this approach for the classification of existing standards.

Communication and Connectivity

Examples of the type of standards that can apply to this KA are:

- Connectivity at physical and link layer
- Network layer
- Service level and application enablers
- Application level API, data models and ontologies
- Management of the protocols

Integration/Interoperability

Examples of the type of standards that can apply to this KA are:

- Profiles
- Certification

Applications

Examples of the type of standards that can apply to this KA are:

- Flexible remote management
- Support methods for installing, starting, updating applications

Infrastructure

Examples of the type of standards that can apply to this KA are:

- Virtualization
- Mobile-edge computing
- Network management
- Network dimensioning, network planning

IoT Architecture

Examples of the type of standards that can apply to this KA are:

- Reference architecture

Devices and Sensor Technology

Examples of the type of standards that can apply to this KA are:

- Device monitoring
- Sensor/actuators virtualization
- Configuration management

Security and Privacy

Examples of the type of standards that can apply to this KA are:

- Communications security and integrity
- Access control
- Authorization, authentication, identity management
- PII (Personally Identifiable Information) management

8.3.4 Status of Standardization in IoT

The IoT has generated the development of standards in order to allow heterogeneous devices to communicate and to leverage common software applications. Interoperability is a great challenge for IoT and standards are the best approach to address it.

Several standardization initiatives currently co-exist, in individual SDOs or in partnerships (e.g. ETSI SmartM2M, ITU-T, ISO, IEC, ISO/IEC JTC 1, oneM2M, W3C, IEEE, OASIS, IETF, etc.) and also in conjunction with a number of industrial initiatives (e.g. AllSeen Alliance, Industrial Internet Consortium (IIC), Open Connectivity Foundation (OCF), Thread protocol, Platform Industrie 4.0, etc.).

In addition to Figure 8.2 where SDO and Alliances were mapped by using two axis representing horizontal and vertical domains, the AIOTI WG03 report on "IoT LSP Standard Framework Concepts" has introduced also two other dimensions: an horizontal axis that represents the market type and a vertical axis that represents the technology that these initiatives cover and focus on. This is displayed in Figure 8.7.

Both of these two figures convey the visual idea of a multiplicity of actors (not to mention the additional multiplicity of standards these actors can produce). It is why the analysis of the IoT standards landscape can give a more precise answer to the question of the proliferation of standards or of the fragmentation of the IoT SDO landscape. This analysis will therefore have several objectives:

- Clarify the issue of SDO landscape fragmentation;
- Identify generic standards that can apply to several vertical domains. These standards can be corresponding to the "horizontal" domain as defined by the AIOTI WG03 but also to cross-domain standards;

Figure 8.7   IoT SDOs and alliances: market and technology.

Source: AIOTI WG03 (IoT Standardisation) – Release 2.6

- Identify standards that are specific to a vertical domain;
- Identify standards gaps.

8.3.5 Overview of IoT Standards Landscape

Regarding IoT landscaping, the work of STF 505 has been focused on the identification and classification of existing standards emanating from SDOs. At the time of this writing, the STF has produced a draft Technical Report than will soon be submitted to public review. In this draft report, two types of standards have been identified:

- "Generic" standards that are common to several vertical domains. A standard is considered as generic when it is used in at least three of the vertical domains;
- "Specific" standards that apply to only one vertical domain.

At this stage, the report (and the standards identified) is under review and cannot be presented globally in this paper. The publication of the final reports of the STF 505 is expected in November 2016, making available the full list of standards that have been analysed.

8.3.5.1 Generic Cross Domain Standards

A number of the standards identified in the IoT landscape are used in several of the analysed vertical domains. Those standards are listed in tables, classified by KAs

- Communication and connectivity
- Integration/interoperability
- Application
- Infrastructure
- IoT architecture
- Device and sensor technology
- Security and privacy

An example of such table with one standard is provided in Table 8.1 where:

- The first column is the short name of the SDO;
- The second column is a short description of the standard;
- The third column provides a description and short analysis of the scope of the standard;

Table 8.1 Example of a generic cross-domain standard

SDO	Standards	Description/Analysis	SC	SL	SF	W	Smo	SE	Sma
Bluetooth	Bluetooth BR/EDR (basic rate/ enhanced data rate) Bluetooth Low Energy (BLE)	Bluetooth is a global wireless communication standard that connects devices. Communication between Bluetooth devices happens over short-range, ad hoc networks known as piconets. The network ranges from two to eight connected devices. When a network is established, one device takes the role of the master while all the other devices act as slaves. Piconets are established dynamically and automatically as Bluetooth devices enter and leave radio proximity. There are different versions of the core specification of Bluetooth. The most common are Bluetooth BR/EDR (basic rate/enhanced data rate) and Bluetooth with low energy functionality. http://www.bluetooth.com/what-is-bluetooth-technology/bluetooth-technology-basics		X	X	X	X	X	

- The seven rightmost columns correspond to the different vertical domains and the meaning of an "X" in a cell is that the Standard(s) described in the row apply to the vertical in the column:
  - SC: Smart Cities
  - SL: Smart Living
  - SF: Smart Farming
  - W: Wearables
  - Smo: Smart Mobility
  - SE: Smart Environment
  - Sma: Smart Manufacturing

The number of currently identified standards per KA and per each vertical domain are displayed in Table 8.2:

As a conclusion, a large number of standards (a total of 70 standards in the above table) may apply to several IoT vertical domains. This set of generic solutions has the potential to provide a common ground to the developers of IoT solutions, irrespective of the specific domain in which they may be applied.

However, this potential will only materialize if the development of IoT standards in vertical domains is making effective use of those standards rather than reinventing similar but not compliant ones, thus increasing the fragmentation of the IoT standards landscape.

8.3.5.2 Domain Specific Standards

Each of the vertical domains has a number of specific standards listed in tables (sometimes empty, in particular when the "generic" standards fully apply), classified by KAs.

An example of such table with one standard is provided in Table 8.3.

Table 8.2 Current STF 505 identified standards per KA and vertical domains

Knowledge Area	3 Vertic.	4 Vertic.	5 Vertic.	6 Vertic.	7 Vertic.	Total
Communication and Connectivity	4	4	4	5		17
Integration/ interoperability	2	3	3	1		9
Application	1	1	1			3
Infrastructure	1	1	2	3	2	9
IoT Architecture		1	3	1	2	6
Device and Sensor technology	1	1		10	4	16
Security and Privacy	1		1	5	3	10

Table 8.3   Example of domain specific standards per vertical domain

SDO	Standards	Description	Analysis
ASHRAE	ISO 16484-5: BACnet – A Data Communication Protocol for Building Automation and Control Networks ISO 16484-6: Method of Test for Conformance to BACnet	BACnet is a data communication protocol for building automation and control.	BACnet provides a standard way of representing the functions of any device, as long as it has these functions, as collections of related information called "objects," each of which has a set of "properties" that further describe it. One of the object's most important properties is its identifier. As devices have common appearances on the network in terms of their objects and properties, messages can manipulate this information in a standard way. It makes possible the interconnection of different vendors' equipment that uses the BACnet protocol http://www.bacnet.org/Overview/index.html

8.3.6 Identifying IoT Standards Gaps

8.3.6.1 Defining Gaps

The work done by the STF 505 on standards identification (described in the previous section) has shown that a number of standards are available, i.e. have reached a final stage (Technical standard (TS) or TR, etc.) in a SDO by the time of writing the report, and can be used for the work of the IoT LSPs.

However, the coverage of the IoT landscape – and the possibility to develop large-scale interoperable solutions – is not fully guaranteed since some elements in this landscape are missing. These missing elements are referred to as "gaps" in this paper. Gaps may also be identified when harmonization or interoperability between a large numbers of potential solutions is missing.

These "gaps" are the main point of interest of the second STF 505 report. Three categories of gaps will be addressed:

- Technology gaps. Some examples in this category are communications paradigms, data models or ontologies, software availability.
- Societal gaps. Some examples in this category are privacy, energy consumption, ease of use.
- Business gaps. Some examples in this category are siloed applications, value chain, and investment.

The identification of gaps has been specially made in view of ensuring that they will be further understood, handled and closed within the IoT community (and possibly beyond).

This identification of gaps relies on an approach that allows for the characterization of gaps, in particular by understanding the type of gaps (see above), the scope of the gap, the difficulties it generates, and other appropriate descriptions.

The STF 505 work does not have the aim to undertake the resolution of the gaps that is left to the proper organizations of the IoT community.

8.3.6.2 Identify Gaps: A User Survey

A critical part of the identification of gaps is the collection of those missing elements. Since they can be of very different nature and may have been detected by very different actors of the IoT community, a mechanism is needed to collect the largest possible information. To this extent, a survey has been built in order to identify as many gaps as possible with the help of the IoT community, in particular the IoT standardization community. The survey aims at:

- Identifying the domain of activity of the respondent;
- Clarifying the respondent objectives and main area of work;
- Defining up to three gaps of all three types described above.

The survey has been largely distributed. At the time of writing this paper, over 190 answers have been collected.

In a second step, these answers have been analysed with the objective to identify commonalities (i.e. related missing functionalities that can be considered as one gap) and associated interoperability frameworks.

8.3.6.3 Example of Gaps

The gaps have been analysed for each of the vertical domains. The sources for the gaps come from both:

- The analysis of the IoT requirements for the vertical domain, based on a variety of sources. These requirements are matched with the list of standards identified in the IoT standards landscaping report and a gaps is identified when no standard seems to match the requirement;
- The analysis of the results of the survey regarding the vertical domain. In this case, the gap has been identified by the respondent and is further analysed.

An example of gaps analysis regarding the Connectivity at Network Layer in the Smart cities domain is shown in Table 8.4.

An example of societal gap identified in the survey in the Smart Cities domain is shown in Table 8.5.

8.3.6.4 Status of Gaps Identification

At this stage, the standard gaps identified are under review and they cannot be presented globally in this paper. The publication of the final reports of the

Table 8.4   Example of gap analysis for connectivity in smart cities

Requirements	Organizations Providing Related Standards
Support of local and remote access to infrastructure services	3GPP, ETSI TETRA, IEEE 802.x, LoRa alliance, ITU, IETF,
Support of point-to-point communications	3GPP, ETSI TETRA, IEEE 802.x, ITU, EnOcean Alliance, DASH7 Alliance,
Support of point-to-multipoint communications	ETSI TETRA, IEEE 802.x, ITU
Support of routing continuity across different network technologies	IETF 6lo
Support of device unique identification	This is a potential gap

Table 8.5   Example of societal gap identified in smart cities by the STF 505 survey

Nature of the Gap	Knowledge Area	Criticality	How Can Standardization or Regulation Improve this?
Privacy and security aspects not sufficiently covered, developed and not real, mature models/solutions seem to be available. This could limit IoT adoption Another social gap is that many decision makers does not have a real understanding of practical potentialities IoT can provide and a dissemination campaign would be useful addressing mainly Public administration.	Communication and Connectivity (network and service levels); Integration/ Interoperability; IoT Architecture; Security and Privacy	3	IoT and big data pose new challenges to an acceptable model of privacy and security management and rules (in terms of civil rights and "industrial privacy/security" guarantees: it is necessary to find out new models/ approaches

STF 505 is expected in November 2016. The full list of the identified standard gaps will then be available.

8.3.7 Conclusions and Further Challenges

The work on IoT standards landscaping and IoT standards gaps analysis is an on-going task that will continue over time. At this stage, the actors involved have established a solid framework and started to collect the results regarding the actual standards and gaps.

The work done by the STF 505 will provide a list of standards and a list of identified gaps. In particular, these gaps will have to be handled, a task that is not in the scope of the STF.

It is expected that it will be the role of AIOTI WG03 and of the IoT LSP and the Coordination and Support Actions (CSA) starting in 2017.

AIOTI WG03 is addressing three key challenges as displayed in Table 8.6 and specific challenges as displayed in Table 8.7 derived from the landscape analysis with collaborations of contributors based on consensus, providing guidance and recommendations on specific technical themes, and around an understanding of how that guidance will be disseminated in an effective way.

This will help to achieve the overall goals as set by the European Commission, and promote the use of open standards that:

- Support the entire value chain,
- Apply within IoT domains and cross-IoT domains.
- Are integrating multiple technologies, based on streamlined international cooperation and which enables easy and fair access to standard essential patents (SEPs).

Table 8.6 Key challenges AIOTI WG03

1	*Architecture*	Guidelines and recommendations, which contribute to the consolidation of architectural frameworks, reference architectures, and architectural styles in the IoT space.
2	*Semantic Interoperability*	Guidelines and recommendations, which contribute to the consolidation of semantic interoperability approaches in the IoT space.
3	*Privacy*	Guidelines and recommendations regarding personal data and personal data protection to the various categories of stakeholders in the IoT space.

Table 8.7   Challenges for AIOTI WG03

Specific Challenges	Timeframe	
	Before 2020	2020 and Beyond
Recommendations of reference architectures, both for experimentation and deployments within IoT domains and cross – IoT domains	X	
Identification of missing (semantic) interoperability standards and technologies within IoT domains and cross – IoT domains and recommendations on solving them	X	
Recommendations and guidelines on solving protocol and interface gaps needed to support new IoT features within IoT domains and cross – IoT domains. In particular, promote the uptake of IoT standards in public procurement to avoid lock-in	X	X
Promoting the use and development of Open Reference Vocabularies and Open Application Programming Interfaces to allow for flexible ad-hoc communication and interaction between different actors within IoT domains and cross – IoT domains	X	X
Provide guidelines on how to translate the Digital Rights Management recommendations within IoT domains and cross – IoT domains	X	
Recommendation of an interoperable IoT numbering space that transcends geographical limits, and an open system for object identification and authentication, which can be applied within IoT domains and cross – IoT domains	X	

Acknowledgments

The content of this chapter is largely relying on the many contributions of the IoT standardization community. The authors would like in particular to thank:

- The reviewers: Georgios Karagiannis, Patrick Wetterwald
- The STF 505 members: Emmanuel Darmois, Joachim Koss, Samir Medjiah, Jumoke Ogunbekun, and Michelle Wetterwald.
- The contributors to the three AIOTI WG03 reports, in particular the report on "IoT LSP Standard Framework Concepts": Howard Benn, Werner Berns, Angel Boveda, Marco Carugi, Pablo Chacin, John Davies, Thierry Demol, Jean-Pierre Desbenoit, Zeta Dooly, Omar Elloumi, Patrick Guillemin, Georgios Karagiannis, Levent Gurgen, Juergen Heiles, Sharadha Kariyawasam, Jochen Kilian, Guenter Kleindl, Paul

Murdock, Thomas Paral, Nigel Rix, Friedhelm Rodermund, Mohammad-Reza Tazari, Martin Serrano, Carlos Ralli Ucendo, Ovidiu Vermesan, Alexander Vey, Patrick Wetterwald.

Bibliography

[1] AIOTI WG03 Report: "IoT LSP Standard Framework Concepts Release 2.0", November 2015. http://bit.ly/1GtzJ5I
Note: the most recent version (Release 2.6 on 30 May 2016) can be found at https://docbox.etsi.org/SmartM2M/Open/AIOTI/

[2] AIOTI WG03 Report: "High Level Architecture (HLA) Release 2.0" November 2015. http://bit.ly/1GtzJ5I
Note: the most recent version (Release 2.1 on 30 May 2016) at https://docbox.etsi.org/SmartM2M/ Open/AIOTI/

[3] AIOTI WG03 Report: "Semantic Interoperability Release 2.0" November 2015. http://bit.ly/1GtzJ5I

[4] STF 505 Draft TR 103 375 "SmartM2M IoT Standards landscape and future evolution", June 2016. https://docbox.etsi.org/SmartM2M/Open/AIOTI/STF505

[5] STF 505 Draft TR 103 376 "SmartM2M; IoT LSP use cases and standards gaps", June 2016. https://docbox.etsi.org/SmartM2M/Open/AIOTI/STF505

9

IoT Platforms Initiative

**Sylvain Kubler[1], Kary Främling[2], Arkady Zaslavsky[3],
Charalampos Doukas[4], Eneko Olivares[5], Giancarlo Fortino[6],
Carlos E. Palau[5], Sergios Soursos[7], Ivana Podnar Žarko[8],
Yiwei Fang[9], Srdjan Krco[10], Christopher Heinz[11],
Christoph Grimm[11], Arne Broering[12], Jelena Mitic[12],
Kathleen Olstedt[13] and Ovidiu Vermesan[14]**

[1]University of Luxembourg, Luxembourg
[2]Aalto University, Finland
[3]Data61, CSIRO, Australia
[4]CREATE-NET, Italy
[5]Universitat Politècnica de Valencia, Spain
[6]Universittà della Calabria, Italy
[7]Intracom SA Telecom Solutions, Greece
[8]University of Zagreb, Croatia
[9]Fujitsu Laboratories of Europe Ltd, UK
[10]DunavNET, Serbia
[11]University of Kaiserslautern, Germany
[12]Siemens, Germany
[13]European Innovation Hub GmbH, Germany
[14]SINTEF, Norway

9.1 Introduction

The scope of Internet of Things (IoT) European Platforms Initiative (IoT-EPI) is to create ecosystems of "Platforms for Connected Smart Objects", integrating the future generations of devices, embedded systems, network technologies, and other evolving ICT advances.

These environments support citizens and businesses for a multiplicity of novel applications. They embed effective and efficient security and

265

privacy mechanisms into devices, architectures, service and network platforms, including characteristics such as openness, dynamic expandability,
interoperability, dependability, cognitive capabilities and distributed decision
making, cost and energy-efficiency, ergonomic and user-friendliness.

Such smart environments are enriched through the deployment of wear-
able/ambulatory hardware to promote seamless environments. The platforms
from the beginning involve and connect technology developers and application developers and complementors who will enhance the impact of the
IoT platform. The IoT European Platforms Initiative is coordinated by two
consortia UNIFY-IoT, Be-IoT and include seven research and innovation
projects.

The IoT-EPI program includes the research and innovation consortia that
are working together to deliver an IoT extended into a web of platforms for
connected devices and objects. The platforms support smart environments,
businesses, services and persons with dynamic and adaptive configuration
capabilities.

The goal is to overcome the fragmentation of vertically oriented closed
systems, architectures and application areas and move towards open systems
and platforms that support multiple applications. The European Commission
funds the IoT-EPI with EUR 50 millions. The projects are presented in the
following sections.

9.1.1 AGILE Project: A Modular Adaptive Gateway for IoT

AGILE builds a modular and adaptive gateway for IoT devices. Modularity
at the hardware level provides support for various wireless and wired
IoT networking technologies (e.g., KNX, Z-Wave, ZigBee, Bluetooth Low
Energy, etc.).

It allows fast prototyping of IoT solutions for various domains (e.g., home
automation, environment monitoring, wearables, etc.).

At the software level, different components enable new features: data
collection and management on the gateway, intuitive interface for device
management, visual workflow editor for creating IoT apps with less coding,
and an IoT marketplace for installing IoT apps locally.

The AGILE software can auto-configure and adapt based on the hardware
configuration so that driver installation and configuration is performed automatically. IoT apps are recommended based on hardware setup, reducing the
gateway setup and development time significantly.

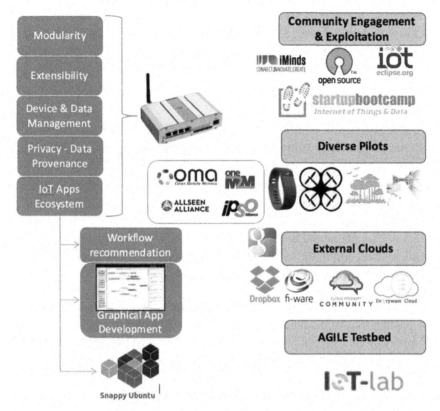

Figure 9.1 The AGILE project overview.

9.1.2 The Challenges

Prototyping an IoT solution is a complex process that involves the careful selection of appropriate components, both at the hardware and software level.

On the hardware side, there are many wireless and wired communication technologies and protocols to choose from and support. For non IP-based networks, there is always a need of a gateway for connecting smart objects to the Internet. Unfortunately, gateways provided by vendors nowadays cannot be extended easily to support new protocols and are domain-specific and closed solutions.

From the software perspective, the plethora of communication protocols complicates the selection of the most appropriate ones for device communication with external services and Machine to Machine (M2M) interaction.

IoT vendors implement their own Cloud-based solutions that are vertical, product-oriented and closed, since there is no standardized way of creating end-to-end IoT applications and no wide acceptance of an IoT platform model. This leads to great privacy and data control issues.

9.1.3 The AGILE Solution

AGILE creates an open, flexible and widely usable IoT solution and puts it at the disposal of industries (start-ups, SMEs, tech companies) and individuals (researchers, makers, entrepreneurs) as a framework that consists of:

- A modular IoT gateway enabling various types of devices (wearables, home appliances, sensors, actuators, etc.) to be connected with each other and to the Internet;
- Data management and device control maximizing security and privacy, at local level and in the cloud, technologies and methodologies to better manage data privacy and ownership in the IoT;
- Support of various open and private clouds;
- Recommender and visual developer's interfaces enabling easy creation of applications to manage connected devices and data;
- Support of mainstream IoT/M2M protocols, and SDKs from different standardization bodies for device discovery and communication;
- Two separate gateway hardware versions: a) the "makers" version, based on the popular RaspberryPi for easy prototyping and b) the "industrial" version for more industrial and production ready environments.

9.1.4 The AGILE Use Cases

To demonstrate the applicability of the AGILE modular hardware and software gateway, the project is developing the following use cases in five different pilots:

- Quantified Self (wearables for self-tracking)
- Crop and livestock monitoring using drones
- Air pollution monitoring
- Port radiation and pollution monitoring using drones
- Smart retail solutions for enhanced shopping experiences.

AGILE will become part of the existing IoT-Lab infrastructure in France managed by INRIA. With more than 2.500 sensors deployed in five locations, AGILE users will have the opportunity to evaluate their IoT applications in real environments, collect and store sensor data, and interact with real devices.

9.2 BIG IoT: Bridging the Interoperability Gap of the IoT

Despite various research and innovation projects working on the IoT, no broadly accepted professional IoT ecosystems exist. The reason for that are high market entry barriers for developers and service providers due to a fragmentation of IoT platforms. Developers who want to make use of smart objects hosted by various providers need to negotiate access to their platforms individually and implement specific adapters. Since the efforts to negotiate individual contracts often outweigh the possible gains, platform providers do not see strong incentives to open their platforms to third parties.

The goal of this project is to overcome these hurdles by **B**ridging the **I**nteroperability **G**ap of the **IoT** (BIG IoT) and by creating marketplaces for service and application providers as well as platform operators.

Previous EC-funded projects that address such enablement of IoT ecosystems are, e.g., IoT-A[1], by providing a common architecture, FIWARE[2] that offers Generic Enablers as building blocks, or projects such as compose[3] and OpenIoT[4], which offer dedicated IoT platforms to aggregate other platforms and systems. BIG IoT will **not** develop **yet another platform** in order to enable cross-platform IoT applications. Instead, we will address the interoperability gap by defining a generic, unified Web API for smart object platforms, called the BIG IoT API. The establishment of a marketplace where platform, application, and service providers can monetize their assets will introduce an incentive to grant access to formerly closed systems and lower market entry barriers for developers.

With this approach based on the generic BIG IoT API, an IoT ecosystem will come to life, as it will offer a functionally rich but at the same time easy way to discover, access, control, manage, and secure smart objects.

The API will be designed in an open community process and the project consortium will engage with current standardization initiatives to receive input and deliver contributions to specifications. The BIG IoT API will be implemented by overall eight smart object platforms.

Following an evolutionary and agile approach, the developed technologies will be concurrently demonstrated in three regional pilots involving partners with strong relation to public authorities. Under a common theme of "smart mobility and smart road infrastructure", various use cases within the pilots will validate the developed technologies.

---

[1] http://www.iot-a.eu
[2] http://www.fiware.org
[3] http://www.compose-project.eu
[4] http://openiot.eu

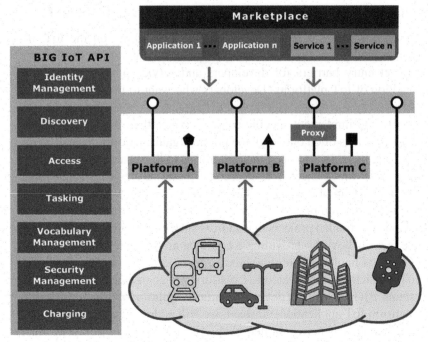

Figure 9.2 Overview of the BIG IoT approach[5].

To foster the external implementation of the BIG IoT API the project will conduct focused dissemination and exploitation activities to leverage the developer community. Further, an Open Call will be conducted as part of the project to engage SMEs in the implementation of the services, applications, and platforms conforming with the BIG IoT.

9.3 bloTope: Building an IoT Open Innovation Ecosystem for Connected Smart Objects

New IoT applications that leverage ubiquitous connectivity, system interoperability and analytics, are enabling Smart City initiatives all over the world. Although the smart city paradigm paves the way for societal and economic opportunities (e.g., to reduce costs for societies, foster a sustainable economic growth, etc.), they also pose architectural and structural issues that must be addressed for businesses to benefit. One of the most critical obstacles is the

---

[5]Icons made by Freepik from www.flaticon.com

vertical silos' model that shapes today's IoT, which hampers developers –
due to the lack of interoperability and openness – to produce new added
value across multiple platforms (data is "siloed" in a unique system, cloud,
domain, and stays there). Several organisations and standardization fora have
understood this critical challenge and started to build up consortia and IoT
initiatives to address it.

The Open Group was among the first ones with the *IoT Work Group*
established in 2010 [1]. More recent initiatives are, for example, the *Web of
Things* initiative at W3C that aims to create open ecosystems based upon open
standards, including identification, discovery and interoperation of services
across platforms; the *Alliance for IoT Innovation* (AIOTI) launched by the EU
with the aim of strengthening links and building new relationships between the
different IoT players (industries, SMEs, start-ups); the *Open Platform 3.0*^{TM} at
The Open Group that focuses more on organization applications and practices;
the *OneM2M global standards* initiative that involves eight standards bodies
for M2M communications; or still the *IEEE IoT* initiative.

Although most of those initiatives promote various types of standards and
specific technology enablers, they all share the same vision about relying
as much as possible on open and interoperable standards to foster open
ecosystems and unlock the commercial potential of the IoT. While in the
US, IoT ecosystems are created around big, multinational players such as
Apple or Google, the EU's strength is rather in smaller and agile companies.
Several past EU initiatives gave rise to a multitude of IoT platforms in various
domains [2] (e.g., OpenIoT cloud platform, BUTLER, FI-WARE, etc.).

Despite these efforts, it is a key challenge for the EU to turn those initial
IoT platforms into economically viable entities and ecosystems. This is the
current focus and goal of the H2020 ICT30 R&I Programme that is composed
of two support action projects and seven R&I projects. In this chapter (in the
next two sections), we briefly introduce the vision, objectives and building
blocks underlying one of these projects named bIoTope (standing for *Building
an IoT OPen innovation Ecosystem for connected smart objects*), along with
a brief overview of the smart city pilots that will be developed.

9.3.1 Building Blocks Underlying the bIoTope Project

This section provides a brief overview of key building blocks that underlie the
bIoTope ecosystem, namely *(i)* IoT standards that will be used as interoperability enablers across various IoT platforms, and *(ii)* context-aware services

(CoaaS) that provide systems with reasoning capabilities that allow them to react appropriately to new situations.

9.3.2 O-MI and O-DF Standards

Primary goal of bIoTope is to enable companies to easily create new IoT systems and rapidly harness available information using advanced Systems-of-Systems (SoS) capabilities for Connected Smart Objects. To this end, bIoTope takes full advantage of messaging standards developed and officially published by The Open Group, notably the Open Messaging Interface (O-MI) and Open Data Format (O-DF) [3, 4] standards.

Those standards emerged out of past EU FP6-FP7 projects, where real-life industrial applications required the collection and management of product instance-level information for many domains involving heavy and personal vehicles, household equipment, etc.

Based on the needs of those real-life applications, and as no existing standards could be identified that would fulfil those requirements without extensive modification or extensions, the partner consortia started the specification of new IoT interoperability standards. O-MI provides a generic Open API for implementing RESTful IoT information system, also using other underlying protocols than HTTP as illustrated in Figure 9.3.

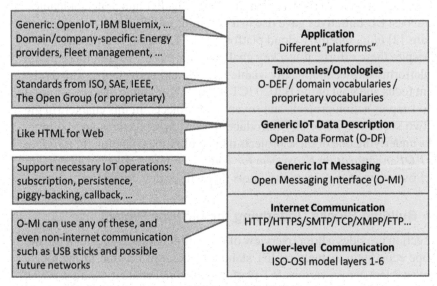

Figure 9.3   Positioning of O-MI and O-DF in IoT protocol stack.

O-DF provides a generic content description model for Objects in the IoT that can be extended with more specific vocabularies (e.g., using domain-specific ontology vocabularies).

In the same way as HTTP can be used for transporting payloads in formats other than HTML, O-MI can be used for transporting payloads also in other formats than O-DF [5].

When used together, O-MI and O-DF provide the necessary tools for "any" IoT information systems to interoperate successfully in ad hoc manners, which is necessary also for dealing with context.

9.3.3 Context-as-a-Service

Context awareness and provisioning (i.e., CoaaS) is a key feature of bIoTope ecosystem. CoaaS will be providing relevant, dependable, trustworthy real-time and historical context to bIoTope services, pilots, platforms and applications through open APIs. Context is defined as "any information that can characterise a situation of an entity" [6]. Tremendous opportunities and challenges exist in implementing and organizing such context-aware systems on different scales, ranging from context-aware printing; context-aware enterprises that respond with agility to an understanding of physical circumstances; context-aware toys that interact with children aware of their age, abilities, parental constraints, context-aware parking areas that tell drivers where to go, to context-aware road intersections that warn drivers of dangerous situations [7].

Context awareness R&D efforts in bIoTope focus on a powerful theoretical framework that enables domain-agnostic representation of context, reasoning about and validation of context. Very little research has been done on context- and situation-prediction [8]. Solid theoretical methods including Particle and Kalman filters, Bayesian Networks, machine learning and Dempster-Shafer theory, Markov models and Reinforcement learning underpin CoaaS. Computationally efficient context fusion from multiple heterogeneous IoT sources is very much a fundamental challenge that is also being addressed in bIoTope. The CoaaS will therefore provide run-time support for advanced context-awareness through context prediction, proactive adaptation, privacy and UI awareness, and personalisation that will lead to the emergence of intelligent, user and object-driven and user-centric services. Our context service components will be open, O-MI/O-DF compliant and, most importantly, scalable.

9.3.4 bIoTope Large-Scale Pilots

Two categories of pilots will be developed in bIoTope to validate the effectiveness of the bIoTope SoS ecosystem for IoT, namely:

- *Domain-specific pilots*: ensure industrial impact through customer networks of bIoTope partners addressing electric car charging stations, self-managing buildings and smart air quality;
- *Cross-domain smart city pilots*: provide proofs-of-concept of IoT system composition and interoperability scenarios in smart city environments (Helsinki, Brussels region, Grand Lyon) including smart metering, shared electric vehicles, smart lighting, etc., as illustrated in Figure 9.4.

9.4 INTER-IoT: Interoperability of Heterogeneous IoT Platforms

INTER-IoT project aims at the design, implementation and experimentation of an open cross-layer framework and associated methodology and tools to enable voluntary interoperability among heterogeneous IoT platforms. The proposal will allow effective and efficient development of adaptive, smart IoT applications and services atop different heterogeneous IoT platforms, spanning single and/or multiple application domains. The project will be tested in two application domains: port transportation and logistics and mobile health; additionally, it will be validated in a cross-domain use case.

Most current existing sensor networks and IoT device deployments work as independent entities of homogenous elements that serve a specific purpose, and are isolated from "the rest of the world". In a few cases where heterogeneous elements are integrated, this is done either at device or network level, and focused mostly on unidirectional gathering of information [9]. A multi-layered approach to integrate heterogeneous IoT devices, networks, platforms, services and data will allow heterogeneous elements to cooperate seamlessly to share information, infrastructures and services as in a homogenous scenario [10].

9.4.1 Open Interoperability

Lack of interoperability causes major technological and business issues such as impossibility to plug non-interoperable IoT devices into heterogeneous IoT platforms, impossibility to develop IoT applications exploiting

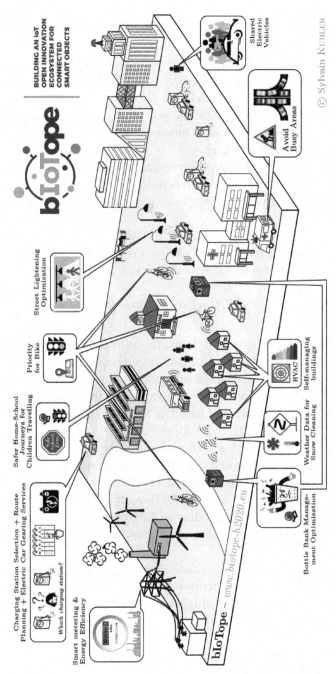

Figure 9.4 Overview of the bIoTope large-scale pilots to be implemented in the bIoTope project.

multiple platforms in homogeneous and/or cross domains, slowness of IoT technology introduction at a large-scale, discouragement in adopting IoT technology, increase of costs, scarce reusability of technical solutions, and user dissatisfaction [11].

The main goal of the INTER-IoT project is to comprehensively address lack of interoperability in the IoT realm by proposing a full-fledged approach facilitating "voluntary interoperability" at any level of IoT platforms and across any IoT application domain, thus guaranteeing a seamless integration of heterogeneous IoT technology [11].

INTER-IoT aims to provide open interoperability, which delivers on the promise of enabling vendors and developers to interact and interoperate, without interfering with anyone's ability to compete by delivering a superior product and experience.

In the absence of global IoT standards, the INTER-IoT project will support and make it easy for any company to design IoT devices, smart objects, and/or services and get them to the market quickly, thus creating new IoT interoperable ecosystems.

The solution adopted by INTER-IoT will include three main products or outcomes (see Figure 9.5):

- **INTER-LAYER**: methods and tools for providing interoperability among and across each layer (virtual gateways/devices, network, middleware, application services, data and semantics) of IoT platforms. Specifically, we will explore real/virtual gateways [13] for device-to-device communication, virtual switches based on SDN for network-to-network interconnection, super middleware for middleware-to-middleware integration, service broker for the orchestration of the service layer and a semantics mediator for data and semantics interoperability [14].
- **INTER-FW**: a global framework (based on an interoperable meta-architecture and meta-data model) for programming and managing interoperable IoT platforms, including an API to access INTER-LAYER components and allow the creation of an ecosystem of IoT applications and services.
- **INTER-METH**: an engineering methodology based on CASE (Computer Aided Software Engineering) tool for systematically driving the integration/interconnection of heterogeneous non-interoperable IoT platforms.

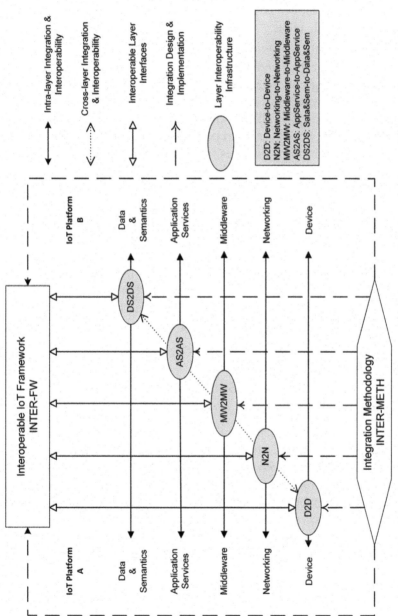

Figure 9.5   INTER-IoT abstract architecture associating INTER-IoT products.

9.4.2 Use-Case Driven

The INTER-IoT approach will be use case-driven, implemented and tested in three realistic large-scale pilots:

- **INTER-LogP**: will be designed and built to specifically accommodate the communication and processing needs of moving vehicles and cargo items (being conceived as moving things according to the IoT paradigm), e.g., by seamless and secure integration of various vehicle telematics solutions as well as mobile devices serving as retrofitting equipment. It will work over smart containers (i.e., reefers and IMOs[6]), trucks and different infrastructures, allowing exchange of information associated with the operations and movements of containers inside the terminal.
- **INTER-Health**: aims at developing an innovative, open integrated m-Health IoT platform for humans monitoring in a decentralized way and in mobility. The integrated platform, derived from existing platforms (i.e., e-Care Tilab platform and BodyCloud)_[15], will be open to be further enhanced by integrating new subsystems by using the INTER-IoT approach.
- **INTER-DOMAIN**: a cross-domain pilot involving IoT platforms from different application domains, including transport and logistics but extendable to other domains.

9.5 symbIoTe: Symbiosis of Smart Objects Across IoT Environments

The IoT is evolving around a plethora of vertical platforms, each specifically suited to a given scenario and often adopting proprietary communications, device and resource control protocols.

The emerging need for cross-domain IoT applications and services highlights the necessity of interoperability across IoT platforms for a unified and secure sharing of and access to sensing/actuating resources.

symbIoTe (Symbiosis of smart objects across IoT environments) steps into this landscape to devise an interoperability framework across existing and future IoT platforms.

---

[6]IMO containers are used to transport safely dangerous goods, available at http://www.imo.org

The framework will enable the discovery and sharing of resources for rapid cross-platform application development and will facilitate the blending of next generation of smart objects with surrounding environments.

9.5.1 The Vision

symbIoTe aims at introducing IoT platform federations, provisioning of domain-specific enablers, sharing of IoT resources and new business models in the IoT landscape.

Vertical IoT solutions focus on specific activities of everyday life, but are restricted to the ecosystem that can be created around a single platform (see "Closed Private" IoT Business models in [16]). Through *federations*, multiple IoT solutions can collaborate so as to i) provide cross-domain solutions, and ii) share IoT resources and the respective measurements in locations originally out of their reach.

For the co-creation of cross-domain solutions it is important that expertise in a certain domain by existing solutions is exploited. To achieve this, current IoT solution providers should wrap and offer their domain-specific platforms in a "Sensing as a Service" manner [17]. This way, important and useful information with respect to a single domain can be provided to third parties, in the form of a *domain-specific enabler*, typically after some pre-processing and aggregation.

To deal with the increasing complexity of IoT systems and reduce the deployment costs, collocated platforms can choose to be cooperative by opening up the access to their resources to third parties and by implementing generic high-level APIs. In addition, they may choose to collaborate by *sharing the common physical resources* in a coordinated way.

Putting the technical details aside, the federations among IoT solution providers need to be supported by the *appropriate business models* in order to be viable. While basic literature on IoT business models is arising [18], the horizontal integration in federations deserves more specific considerations since the current IoT value chain includes more stakeholders like infrastructure providers, IoT platform providers, Cloud operators, ISPs and application developers.

9.5.2 The Technical Approach

symbIoTe builds around a hierarchical IoT stack connecting smart objects and IoT gateways within smart spaces with the Cloud. Smart spaces share the available local resources, while platform services running in the Cloud should enable federations and open up northbound interfaces to third parties. The architecture comprises four layered domains, as depicted in Figure 9.6.

The *Application Domain* offers a high-level API for a unified view on different platforms to enable cooperation and support cross-platform discovery and management of IoT resources, as well as data acquisition and actuation in accordance with platform-specific business rules.

The *Cloud Domain* hosts the Cloud-based building blocks of specific platforms. To enable platform federations and sharing of resources, a symbIoTe interworking interface will be defined and implemented for the exchange of information between two collaborating IoT platforms.

The *Smart Space Domain* comprises smart objects, IoT gateways as well as local computing and storage resources. To enable dynamic sensor discovery and configuration, as well as dynamic sharing of the wireless medium, symbIoTe introduces the symbIoTe middleware, which will expose a standardized API for resource discovery and configuration within a smart space, and implement a sensor-discovery protocol for a simplified integration of sensors with platforms hosted in particular smart space domains.

The *Device Domain* spans over heterogeneous devices, which should be capable to dynamically blend with the surrounding environment and get discovered by the symbIoTe middleware. Smart objects can be configured on the fly to be integrated with different IoT platforms hosted within the smart space, preventing the lock-in to specific IoT platforms.

9.5.3 The Use Cases

symbIoTe use cases are targeting typical daily environments to assist people seamlessly while performing their daily activities. The diversity of the considered environments is ideal to showcase platform interoperability.

Smart residence will enable automatic discovery and configuration of devices in homes and offices as well as sharing of available resources between collocated platforms. The basic idea is to exploit local resources and dynamic service composition to manage and access functions across any available device.

Smart campus will develop campus-wide smart services across various platforms with a focus on collaboration services which utilize indoor navigation and room/equipment booking. In addition, it will enable "eduroam-like" IoT services for visiting students and staff and showcase device roaming across IoT domains.

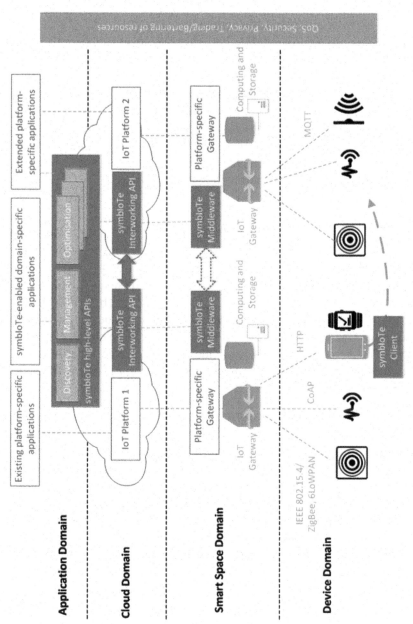

Figure 9.6   The symbIoTe high-level architecture.

Smart stadium will enable indoor location services while supporting strict security and privacy policies. The goal is to link digital and physical worlds so as to create a unique experience for stadium visitors.

Smart Mobility and Ecological Routing will bring together existing city-wide air quality measurement infrastructures with wearable air quality sensors to predict the total emission levels commuters are exposed to. A domain specific enabler will offer a service for the calculation of the ecologically preferable routes for motorists, bicyclists and pedestrians.

Smart Yachting will automate the information processes between a boat and the mainland, to allow i) users on a boat to identify automatically the territorial services and ii) the port authorities to automatically send various land information to the boat, e.g., during the mooring phase.

9.6 TagItSmart

The IoT is about connecting objects, things, and devices, billions of them. What is still out of reach due to technological limitations and the cost of deployment are mass-market products: a carton of milk, a package of steak, a basket of apples, a book, a CD etc. Today, these products are identified by printed tags (such as barcodes or QR codes). These codes relate to the product they tag, not to the unique unit/object that holds the tag. Once attached to an object, tags are usually static and the information they provide does not change, regardless of the state or events happening in the immediate environment of that product.

9.6.1 Vision

Leveraging the features of functional codes (such as QR codes printed using functional inks) to change according to the context changes of each tagged product together with wide availability of smart phones that can capture/record/transmit these codes we can create context sensors for mass market products and convert mass market products into connected mass market products with unique identity that can report on their environment.

This opens up possibilities for a completely new range of services to be created and consumed by the user, and for the user. The outcome will be the creation of an almost infrastructure-less IoT framework applicable in multiple industry sectors.

Funcational ink + optical tags + crowd sourced smartphone

= IoT for mass market products

9.6.2 Objectives

The overall objective of TagItSmart is to create a set of tools and enabling technologies integrated into a platform with open interfaces enabling users across the value chain to fully exploit the power of condition-dependent functional codes to connect mass-market products with the digital world across multiple application sectors.

TagItSmart will define a framework, enabling technologies and the tools required to design and exploit functional codes across multiple application sectors in a secure and reliable manner. The project will leverage clearly identified and well-established catalysts (i.e., functional inks, printed circuit NFC, smartphones pervasiveness and cloud computing) to enable inclusion of any mass market product into the world of connected objects.

Functional inks and printed NFCs will be used to create functional codes, which will provide sensing capabilities to the objects they are attached to. Product manufacturers, shopping centres, supply chain providers and other stakeholders from different sectors will be able to leverage the framework to easily and automatically produce and deploy these codes according to their needs and the properties they need to observe and track. Functional

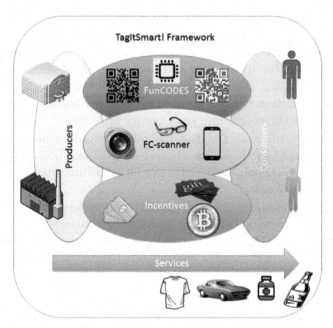

Figure 9.7 TagItSmart concept overview.

codes scanners (fixed and provided by existing infrastructure, or supported by participatory engagement of consumers) will be used to obtain data from functional codes throughout the product lifecycle.

9.6.3 The Approach

The following are the main characteristics of the TagItSmart project's approach and methodology:

1. Leverage and re-use the existing solutions, components, architectures, communities, ecosystems, do not build from scratch.
2. Build open systems, use open source and interact with other relevant initiatives and ecosystems identified in collaboration with the community.
3. Start building a community and an ecosystem immediately from the beginning of the project.
4. Use co-creation methodology to specify new use cases, listen and be agile to react on the input.
5. Pilot solutions early. Involve community in evaluation. Iterate.
6. Use open calls to address new use cases and extend TagItSmart Functionality.
7. Aim to create commercial opportunities already during the project.

9.6.4 Industry Impact

Consumers will be able to get additional assurance about authenticity of the product, information about the way the item in hands was handled, transported etc. as well as to receive other services and incentives provided by manufacturers, retailers and third party service providers.

Manufacturers will be able to track individual items they manufactured throughout the product lifetime, thus offering insights into how the item is being handled, used and finally disposed of as well as enabling them to interact directly with the consumers.

Retailers will be able to monitor individual items on the shelves, improve logistics and supply chain and offer new services to consumers.

Developers will be able to build applications and services on top of the open API and provide them to consumers whenever a tagged product is scanned.

The TagItSmart project includes two open calls for the third parties willing to build new components for TagItSmart framework as well as new services leveraging the provided open API. For this activity, a budget of €1.2 million has been allocated.

9.6.5 Use Cases

The TagItSmart project plans to address at least the following 5 use cases.

1. **"Digital Beer"**: Digital beer is a product that is marked with functional codes, during its production and distribution using the TagItSmart platform. For the digital beer, the TagItSmart enables item-level control, lifecycle management, digital engagement and authenticity control. The functional codes' sensing capabilities make it possible to track products internal and external conditions throughout the lifecycle.

2. **"Lifecycle and Consumer Engagement"**: This use case will implement a system that allows full lifecycle management of every fast-moving consumer good (FMCG) that motivates and helps companies and citizens to recycle their waste items, overcoming the current limits.

3. **"Brand protection"**: Brand protection is a serious issue that matters to manufacturers of both high value products such as high fashion textiles and accessories, and low value products manufacturers like Aspirin. This use case reveals a simple but powerful method for authenticity checking that could act either as a stand-alone security platform or as an added feature to other TagItSmart use cases.

4. **"Supply Chain and Dynamic Pricing"**: This use case provides consumers with the capacity to check the time elapsed from when the product is packaged, and the conditions in which it is stored and transported to the shelf, on top of other basic product information. The products can at the same time be priced dynamically reflecting the goods condition, eliminating the end consumers' doubt about the quality or the price of a product.

5. **"Home Service"**: This is a use case for a new business model. By moving the value downstream along the supply chain, the retailer acts as a service provider or as a trusted 3rd party. Additional services on top of the traditional after-sale services can be provided to customers, making them better enjoy the services and products, improving their satisfaction level.

9.7 Vicinity

VICINITY – "Interoperability as a Service for the IoT: a bottom-up approach" [20]

The VICINITY project will build and demonstrate a platform and ecosystem for IoT infrastructures that will offer "Interoperability as a Service".

The platform aims to be device and standard agnostic, and will rely on a decentralised and user-centric approach. VICINITY aims to retain full control of the ownership and distribution of data across the different IoT domains.

VICINITY introduced the concept of virtual neighbourhood, where users can share the access to their smart objects without losing the control over them. A virtual neighbourhood will be a part of an IoT infrastructure that offers decentralised interoperability and will release the vendor locks that are present in the current IoT ecosystems.

New independent value added services across IoT domains may benefit from the availability of the vast amount of data in semantic formats that are generated by IoT assets.

9.7.1 Challenges

The lack of integration across different disciplines, vendors and standards prevents exploitation of the huge potential in successful large-scale IoT implementations.

It is difficult to control the data flow and privacy settings within a virtual neighbourhood consisting of IoT devices, and it creates both social and technological barriers, which affects the development of new value-added services.

Identifying, configuring, managing and updating information concerning the IoT ecosystem demands technical expertise, which makes it less feasible for the smaller stakeholders, and ultimately may lead to slow adoption rate among the users that may be in the most need – especially within the eHealth and assisted living domain. This is however also something that influence smart home appliances and green energy implementations, as well as how smart home systems are tied in with transportation and the nearby surroundings.

9.7.2 VICINITY Solution

VICINITY presents a virtual neighbourhood concept. A decentralized approach resembling a social network will be used. The users are allowed to configure installations and integrate standards according to the preferred services, as well as being able to fully control their desired level of privacy.

Data exchange between different devices is handled through the VICINITY open interoperability gateway, which reduce the need for having a technical background in order to exploit to the VICINITY ecosystem.

An API will allow for easy development of an adapter to the platform. Once an IoT infrastructure is integrated, its owner can simply manage the access to his/her IoT data and controls using the VICINITY neighbourhood manager (VNM).

Connecting to detected IoT infrastructures is handled by the open VICINITY auto discovery device. The device will automatically discover the smart objects. These devices will appear in a device catalogue, and will allow the users to manage access rules for the discovered smart objects.

9.7.3 Demonstration and Impact

VICINITY will provide an IoT platform that can connect islands that were previously isolated, and will allow integration of end-users and creation of new business models. VICINITY will pave the way for large-scale demonstration of the applicability of the solution in different use cases that implement and demonstrate different value-added services on top of the VICINITY platform.

The first use case will be a smart energy micro-grid that is enabled by municipal buildings (Enercoutim, PT). The VICINITY value-added services will provide users with information on potential energy savings and thereby increase awareness of the contributors.

The second use case will show how to combine infrastructure from different domains: a Smart Grid ecosystem will be combined with an Assisted Living use case (Tiny Mesh, NO).

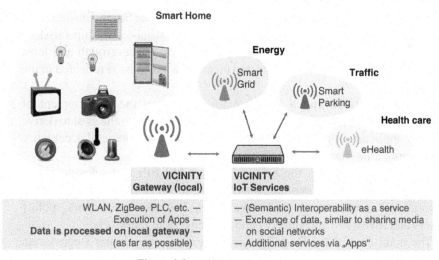

Figure 9.8 VICINITY overview.

The third use case will be eHealth (GNOMON, GR) which looks at the specific needs and constraints of eHealth. Value-added services will include the detection of abnormal events, and the possible finding and clustering of similar patients based on data mining.

The final use case will show how a large number of data sources from different domains can be combined in an intelligent parking application (Hafenstrom, NO). This will use data from booking, heating management and health status while considering how users can be incentivised to use the application.

VICINTY is open and welcomes the participation of further interested consumers, integrators and developers of value-added services.

9.8 Be-IoT

The vision of Be-IoT is to build a broad and vibrant ecosystem for the overall project IoT-EPI increases the collaboration of the research and innovation projects within the overall initiative, generates economic impact through new innovative business models and creates trust in the IoT by transparent information about societal challenges such as privacy and security implications.

SMEs have been set to take on a very important role as a focus group in Europe, since SMEs are at the heart of innovation in the economy. They play a vital role with their capacities to generate new ideas and quickly transform these into business. Their importance is illustrated in the Small Business Act (SBA) and also reflected in the Horizon2020 industrial leadership mission, which states that Europe needs innovative SMEs to create growth and jobs. Their importance and needed support to create new business is reflected in the Be-IoT project.

Be-IoT is establishing a structure for supporting the development of standardized IoT technologies and disseminating those with the goal to derive use case applications and business models and to create societal acceptance of IoT applications across Europe.

Be-IoT project builds the bridge between IoT-EPI and relevant stake-holders (e.g., potential customers such as European SMEs as well as larger corporations, entrepreneurs and developers, but also researchers, policy makers and investors) and thus expand the ecosystem massively.

The main goal of the Be-IoT project is to build an adopter's ecosystem focussing on developers, entrepreneurs and end-users.

Innovation activities such as idea challenges, business model challenges and hackathons create awareness and make sure a variety of end-user interests will be taken into account when building the platforms.

Idea challenges and business model challenges derive specific business opportunities. Best practices and use cases will be derived from these and implementation will be pursued, in collaboration with both investors and SMEs.

The project will work on stimulating the platform adoption by hosting idea and business challenges and hackathons, while setting the ground for upcoming business building activities by creating awareness and also by facilitating and fostering societal acceptance (e.g., by running a variety of innovation, communication to the businesses and to the public and dissemination activities).

9.9 UNIFY-IoT

UNIFY-IoT objectives are to stimulate the collaboration between IoT projects, between the potential IoT platforms and support these in sustaining the IoT ecosystems developed by focusing on complementary actions, e.g., fostering and stimulating acceptance of IoT technology as well as the means to understand and overcome obstacles for deployment and value creation.

UNIFY-IoT is the "working partner" of the Alliance for IoT Innovation (AIOTI) and the IoT European Research Cluster (IERC) by coordinating and supporting the activities on innovation ecosystems, IoT standardisation, policy issues, research and innovation.

The overall concept underpinning UNIFY-IoT is to stimulate strategic cooperation and cross project support between the projects and potential

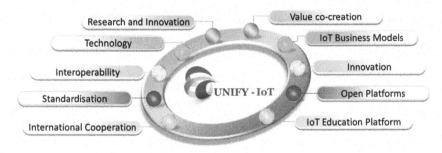

Figure 9.9 UNIFY-IoT activities overview.

platforms that will be used, conceived and developed under IoT-EPI. UNIFY-IoT aim is to:

- Identify new research and innovation mechanisms
- Derive joint exploitation strategies on how to make successful ecosystems emerge;
- Involve and coordinate the cooperation with the AIOTI, ECSEL JTI, cyber-physical stakeholders;
- Give input on and support extend the international cooperation
- Respond to the societal challenges for Europe.

The main activities are focusing on:

- **Value co-creation** – bringing together the various stakeholders in the IoT ecosystem to work towards a mutually agreed outcome using IoT interoperable solutions and evaluate the value co-creation by analysing the results of the projects.
- **IoT Business Models** – surveying and analysing existing business models related to IoT: from specific deployment in case of process optimisation in a company, to, at the opposite, providing a technological element to the open markets, and produce a taxonomy of business models.
- **Innovation Support** – analysing existing IoT platform deployments and analyses at the innovation and other activities of those deployments. It assesses the relative success of the platform adoptions and identifies common innovation activities in the most successful platforms.
- **IoT Open platforms concepts** – building upon on the open platforms activity chain started by the IERC an combine it with other initiative documenting project outcomes.
- **IoT Education platform** – interacting with stakeholders to identify opportunities for interaction between IoT platforms and education institutions to ensure that future graduates are conversant with emerging IoT platforms and the opportunities they present. UNIFY-IoT is leveraging the knowledge process supporting the emergence of an IoT Curricula and education platform in Europe.
- **Standardisation Support** – sensing the global trends in term of interoperability and de-facto standards, and interacts with standardisation bodies including ETSI and CEN/CENELEC to systematise de facto standards emerging from the IoT-EPI projects. The project is cooperating closely with the working group on standardisation of the AIOTI to ensure a coordinated approach to standardisation.

Bibliography

[1] The Open Group. An Introduction to Internet of Things (IoT) and Lifecycle Management: Maximizing Boundaryless Information Flow^{TM} through Whole-of-Life Lifecycle Management Across IoT, https://www2.opengroup.org/ogsys/catalog/W167, Accessed 24 May 2016.

[2] O. Vermesan and P. Friess. *Internet of Things – From Research and Innovation to Market Deployment*. River Publishers, 2014.

[3] The Open Group, October 2014. Open Messaging Interface Technical Standard (O-MI), https://www2.opengroup.org/ogsys/catalog/C14B, Accessed 1 June 2016.

[4] The Open Group, October 2014. Open Data Format Technical Standard (O-DF), https://www2.opengroup.org/ogsys/catalog/C14A, Accessed 1 June 2016.

[5] K. Främling, S. Kubler, A. Buda. Universal Messaging Standards for the IoT from a Lifecycle Management Perspective. *IEEE Internet of Things Journal*, 1(4):319–327, 2014.

[6] G. D. Abowd, A. K. Dey, P. J. Brown, N. Davies, M. Smith and P. Steggles, Towards a better understanding of context and context-awareness, in *Handheld and Ubiquitous Computing*, 1999, pp. 304–307.

[7] Perera, C., Zaslavsky, A., Christen, P. and Georgakopoulos, D. Context Aware Computing for the Internet of Things: A Survey, *IEEE Communications Surveys*, 16(1), 414–454, 2014.

[8] Boytsov, A. and Zaslavsky, A. A Formal Verification of Context and Situation Models in Pervasive Computing, *Pervasive and mobile computing*, 9(1), 98–117, 2013.

[9] Shancang Li, Li Da Xu and Shanshan Zhao, The Internet of Things: A survey, *Computer Networks*, 54(15): 2787–2805, 2015.

[10] Miao Wu, Ting-Jie Lu, Fei-Yang Ling, Jing Sun, and Hue Du, Research on the architecture of Internet of Things, *In Proceedings of the 3rd IEEE International Conference on Advanced Computer Theory and Engineering*, Chengdu (PRC) 2010.

[11] Pablo Giménez, Benjamin Molina, Jaime Calvo, Manuel Esteve and Carlos E. Palau, I3WSN: Industrial Intelligent Wireless Sensor Networks for indoor environments, *Computers in Industry*, 65: 187–199, 2014.

[12] M. Ben Alaya, S. Medjiah, T. Monteil, and K. Drira, Toward Semantic Interoperability in oneM2M Architecture, *IEEE Communications Magazine*, 53(12): 35–41, 2015.

[13] Giancarlo Fortino, Daniele Parisi, Vincenzo Pirrone and Giuseppe Di Fatta, BodyCloud: A SaaS approach for community Body Sensor Networks. *Future Generation Comp. Syst.* 35: 62–79, 2014.

[14] Juan V. Pradilla, Carlos E. Palau and Manuel Esteve, Lightweight Sensor Observation Service (SOS) for Internet of Thing (IoT), In *Proceedings of ITU Kaleidoscope Conference*, Barcelona (Spain), 2015.

[15] Gianluca Aloi, Giuseppe Caliciuri, Giancarlo Fortino, Raffaele Gravina, Pasquale Pace, Wilma Russo and Claudio Savaglio, A Mobile Multi-Technology Gateway to Enable IoT Interoperability, In *Proceeding of the IEEE IoTDI Conference*, Berlin (Germany) 2016.

[16] S. Leminen, M. Westerlund, M. Rajahonka, R. Siuruainen, S. Andreev, S. Balandin, Y. Koucheryavy, *Towards IoT ecosystems and business models*, in: Internet of Things, Smart Spaces, and Next Generation Networking, Springer Berlin Heidelberg, 2012, 15–26.

[17] J. Soldatos, N. Kefalakis, M, Serrano, M. Hauswirth, *Design principles for utility-driven services and cloud-based computing modelling for the Internet of Things*, in: International Journal of Web and Grid Services (IJWGS) 10.2/3 (2014), pp. 139–167.

[18] M. Westerlund, S. Leminen, M. Rajahonka, *Designing business models for the internet of things*, Technology Innovation Management Review, 2014, 4. Jg., Nr. 7, S. 5.

[19] McKinsey & Company.

[20] *VICINITY project website – http://vicinity2020.eu/*

[21] *Cliparts taken from:* https://openclipart.org/

10

European IoT International Cooperation in Research and Innovation

Philippe Cousin[1], Pedro Maló[2], Congduc Pham[3], Xiaohui Yu[4], Jun Li[4], JaeSeung Song[5], Ousmane Thiare[6], Amadou Daffe[7], Sergio Kofuji[8], Gabriel Marão[8], José Amazonas[9], Levent Gürgen[10], Takuro Yonezawa[11], Toyokazu Akiyama[12], Martino Maggio[13], Klaus Moessner[14], Yutaka Miyake[15], Ovidiu Vermesan[16], Franck Le Gall[1] and Bruno Almeida[17]

[1]Easy Global Market, France
[2]FCT NOVA & UNINOVA, Portugal
[3]Université de Pau et des Pays de l'Adour, France
[4]China Academy of Information and Communications Technology, China
[5]Sejong University, South Korea
[6]Université Gaston Berger, Sénégal
[7]Coders4Africa, Senegal/Kenya/USA
[8]Brazilian IoT Forum, Brazil
[9]Universidade de São Paulo, Brazil
[10]CEA, France
[11]Keio University, Japan
[12]Kyoto Sangyo University, Japan
[13]Engineering Ingegneria Informatica Spa, Italy
[14]University of Surrey, UK
[15]KDDI R&D Labs, Japan
[16]SINTEF, Norway
[17]UNPARALLEL, Portugal

The IoT is now a global happening that is requiring cooperation at international level to address its key challenges. Europe has established as a priority the international cooperation on IoT research and innovation. The work revolves around aligning strategies and plans for IoT globalisation but also exploring differentiations, and specificities for local exploitation of IoT. Notice: EU is cooperating with African countries on cost-effective open IoT innovation; Europe is supporting Brazil to build-up its IoT ecosystem supported on EU best practices; the EU-China IoT Advisory Group is active on pushing global IoT standards while developing competitive IoT solutions; the EU-Japan joint cooperation follows-on on the integration/federation of IoT with Big Data and Cloud; and the EU-Korea engagement is looking at major global IoT standardisation activities; EU-US cooperation is active especially via the respective global IoT initiative frameworks, the AIOTI and IIC. And cooperation is expected to start with India on the vision of a connected and smart IoT based system for their economy, society, environment and global needs. This chapter reports on EU international cooperation activities with partner countries and regions on the Internet of Things (IoT).

10.1 Introduction

The importance of international cooperation in science and technology is explicitly recognised in the European Union's Innovation Union flagship initiative and the projects for Horizon 2020, the EU funding programme for research and innovation. On September 14th, 2012, the European Commission set out its new approach to international cooperation under Horizon 2020 in a communication entitled "Enhancing and focusing EU international cooperation in research and innovation: a strategic approach" [1]. In-line with this approach, international cooperation activities developed under Horizon 2020 should contribute to the objectives of:

- Strengthening EU excellence and attractiveness in research and innovation and its economic and industrial competitiveness;
- Tackling global societal challenges; and
- Supporting the Union's external policies.

The Commission's Communication document calls for a systematic and coherent identification of priorities for international cooperation with the EU's partner countries and regions, with a view to subsequently implementing these through activities with the necessary scale and scope, in particular in the context of Horizon 2020. The Communication equally stresses that this strategic priority setting exercise should fully reflect the state of play in the policy dialogues between the EU and its partner countries.

To ensure that international cooperation activities are developed on the basis of common interest and mutual benefit and create win-win situations, the Communication offers four criteria for guiding the identification process. International cooperation adds value when:

- Synergies and complementarities can be created in research and innovation capacity;
- There are opportunities for access to new or emerging markets;
- The activities contribute to meeting the EU's international commitments, as e.g., in the Millennium Development Goals;
- There are adequate legal and administrative frameworks in place to engage in cooperation, also including lessons learnt from previous cooperation.

The Communication also calls on this priority setting process to be reflected in multi-annual road maps for international cooperation with its key partner countries and regions. The road maps for international cooperation, which are included in a Staff Working Document [2], provide examples of the outcome of this priority setting exercise. For each of the partner countries and regions, they provide a full overview of the framework governing the cooperation and the current state of play as regards the cooperation, including information on the way this has been addressed in the first Horizon 2020 work programmes. Most importantly, they provide an overview of what are considered to be the priorities for future cooperation (using a medium term perspective) with the partner in question, reflecting the current state of agreement in the policy dialogue.

As far as International cooperation in IoT is concerned, we can recall on important cooperation aspects as presented above:

- Need to think global for tackling global societal challenges;
- Need to identify synergies and complementarities that can be created in regards to research and innovation capacity;
- Need to look at economic dimension and business opportunities for strengthening EU's excellence and attractiveness in research and innovation and its economic and industrial competitiveness; opportunities to access new or emerging markets.

The cooperation with countries presented in this chapter highlights either important cooperation actions leading to get global solutions (e.g., on standardisation, governance, privacy) or important differentiation to address new markets (e.g., affordable solutions for Africa, new IoT solutions for Brazil, etc.).

10.2 IoT in South Korea and Cooperation with EU

South Korea is known early adopter of new technology and ranked in IDC's 'IoT Index' for 2013. South Korea government has established the 'Mid- and Long-term R&D plan for IoT' that links existing R&D projects classified into parts of the entire ecosystem. Based on the roadmap, many activities and projects are occurring across South Korea. While South Korean's government is helping collaboration between companies, research institutions and universities, individual companies and developers are contributing the entire IoT ecosystem. Also there are strong R&D cooperation between the private sector and the military, which is expected to contribute to advancement of the military applications, and improve leadership in international standards through joint research with major countries including the EU[1]. A recent IoT related R&D direction in South Korea is moving towards integrating AI technologies to IoT in order to support intelligent IoT services. Here, various IoT related activities fostering the IoT ecosystem in Korea are described.

10.2.1 Open Innovation and Open Platform

One of the key criteria that the South Korean government highlights in strategy and planning is 'Open Innovation and Open Platform'. Different from traditional industry, IoT has a characteristic that anyone interested can develop and provide services using global standards based open platforms. In such environment, ideas can easily be developed into new services, and the potential of each individual can be maximised.

In order to launch IoT services quickly, there needs to exist an ecosystem so that developers and users can actively participate and share their technologies and experiences. Open Alliance for IoT Standard (OCEAN) is an example of open innovation and platform to foster the IoT industry supported by the government. OCEAN is a consortium sharing open source and software products that are developed based on the IoT international standard oneM2M to help enhance coordination between companies and help develop the IoT industry. The consortium started with some 50 firms and institutions but now has more than 350 firms worldwide. The IoT platform distributed via OCEAN had been developed through government support.

10.2.2 Large-Scaled IoT Pilot Projects

In order to foster the deployment of IoT solutions in South Korea, the South Korean government have identified several areas including smart city and

---

[1]EU and Korea reaffirmed to strengthen the agreement of the Nov'2013 summit, where both sides agreed on promoting R&D collaboration in the area of ICT including the IoT.

daily healthcare. The government selected these areas based on their large influence to on people's daily life.

In the case of IoT-enabled Smart Cities, Busan, which is the second largest city in South Korea (3.6 M population), teams up with a consortium that comprises of industry and academic members in order to foster an ecosystem for smart city industry and support for Korea's small- and medium-sized companies in various sectors, e.g., social security, transportation, energy efficiency and urban life. A main purpose of the Busan smart city is to establish an open smart city platform based on a global IoT/M2M standards (i.e., oneM2M) and implement an IoT enabled test bed in Busan. The city is designed to guarantee an interoperability between S (Service) – P (Platform) – N (Network) – D (Device) – Se (Security) ecosystem and meet global standard to prepare global City-to-City interoperability and enable expansion to the ecosystem in Busan intends to provide public information about infrastructure, transportation, security and safety, so that new services and technologies using such information can be boosted. The planned services included smart streetlights, a lost child prevention system, smart parking and a building energy management system.

The government also supports a similar large-scaled IoT pilot project together with the city of Daegu focusing on daily healthcare. In order to make sure all IoT data, devices and services are interoperable, all IoT enabled large scaled projects are recommended to use the same global IoT/M2M service layer standards, i.e., oneM2M. In this way, these large scaled IoT test beds can guarantee sustainability even after finishing project periods.

10.2.3 Global Collaboration

The IoT is commonly recognised as a fundamental game changing technology across many industrial sectors and social solutions. Many experts and studies agree that the biggest challenge for the IoT is to overcome market fragmentation and to achieve interoperability between many established silos and global IoT platforms. Therefore, South Korea government strongly supports global collaboration. A new project called WISE-IoT has been started as part of jointly funded R&D programs between South Korea and EU. As shown in table Table 10.1, the project is a joint endeavour with leading IoT companies, research institutes and universities to provide global semantic interoperability.

WISE-IoT has a plan to extend existing IoT reference architectures for achieving interoperability and interworking while strengthening oneM2M deployments. The project started from the existing test beds and experimental

Table 10.1   WISE-IoT EU-Korea joint IoT project members

European Union	South Korea
Easy Global Market	Sejong University
NEC Europe Ltd.	Korea Electronics Technology Institute
Telefonica	SK Telecom
Commissariat à l'énergieatomique et aux énergies alternatives	Korea Advanced Institute of Science and Technology
University of Cantabria	Samsung SDS
Liverpool John Moores University	Kyungpook National University
Telecom Sud Paris	Axstone
Ayuntamiento de Santander	Gimpo Big Data
University of Applied Sciences and Arts Northwestern Switzerland	Gangneung Science Industry Foundation
	IreIS

systems built upon current reference implementations for the various IoT systems, e.g., oneM2M systems in Korea, FIWARE in Europe, LoRa in both regions, and various local IoT technologies such as OCF and AllSeen. These solutions will be made interoperable by semantic annotation of the basic data and a specific reasoner with the knowledge for using semantic information.

The economies in Europe and South Korea are high-tech, knowledge-based societies that are selling products and services across the globe. WISE-IoT aims to enable new business in which essential features like information analytics, intelligent decision making, and reliable execution of workflows and processes can remain in the control of the knowledge workers, while services can be quickly applied to any new IoT data lake as it becomes available in a new city or factory. The outcome of the project will help to establish new global value chains. The cooperation of South Korea and EU is essential to lead the way to the global IoT services and new value chains around the world.

Apart from the WISE-IoT project, South Korea also established a "Global Council of Public and Private sectors for IoT" and the 'IoT Innovation Center' to improve partnerships between software, device, or user businesses and large businesses/SMEs. This scheme aims to foster small yet strong IoT businesses for global expansion by providing education for creative entrepreneurship and conducting projects in teams of large companies and SMEs.

10.3 Global IoT Challenges Seen from China

Note: For five years, EU and China experts met about twice a year to discuss cooperation on IoT. The result of discussion is now available in a white paper on EU-China cooperation for IoT [3]. The summary below is based on the content of the white paper.

In China, the IoT has become an important carrier for strategic information industries and integrated innovation. The Central Government and local authorities have consistently attached great importance to IoT through the Inter-Ministerial Joint Conference, the tenth action plans for the IoT development and the annual special fund for IoT development giving substantial support for industrial development. As a result, China's IoT development shows now a strong momentum of development. In 2014, China's IoT industry scale expanded beyond 620 billion yuan, with a year-on-year growth of 24% [22]. The M2M terminals in China exceeded 73 million units, with a year-on-year growth of 46%, accounting for 30% of the global total [4]. Beijing-Tianjin, Shanghai-Wuxi, Shenzhen-Guangzhou, and Chongqing-Chengdu form the four core industry clusters with their unique features, where a number of leading enterprises have emerged. Moreover, IoT third party operation service platforms are rising in traffic, security, health care, IoV, energy-saving areas, and IoTaaS.

10.3.1 China Policy on IoT

China's IoT policies emphasise on demonstration and cluster effects, and the policy environment will continue to be improved from the top design, organisational mechanisms, think-tank support and other fields of activities:

- Planning documents pointing out directions for development of stages: following the 12th Five-Year-Plan for IoT development, China's State Council issued the Guidance on Advancing Orderly and Healthy IoT Development, which further clarified the goals, ideas and areas of focus of China's IoT development;
- Establishment of the Inter-Ministerial Joint Conference system and the Expert Consultation Committee for IoT development: the NDRE, the MIIT, and the most coordinated for the top-level design of the IoT development and promoted IoT development in China;
- Formulation of ten action plans for IoT development: the plans cover various perspectives, including top-level design, standard development, technology development, application and promotion, industry support, business models, safety and security, supportive measures, laws and regulations, personnel trainings, etc.;
- Financial support such as the special fund for IoT development: the annual support directions of the special fund are set against the

development demands from key IoT technology R&D projects and IoT systems development projects in key are as during the year; additionally, the annual support measures will to be adjusted and optimised.

10.3.2 IoT Applications in China

From a macroscopic perspective, China's IoT application development presents two typical types: "focus-oriented objective" and "overall and global covering objective". The "focus-oriented objective" are IoT applications in specific industries: CPS (Cyber-Physical Systems) for mapping virtual models to the real world, while IoT is the core of CPS.

In the field of industrial manufacturing, IoT has been widely applied in intelligent equipment management, environmental real-time monitoring, materials/product tracing and other areas. The applications of CPS will enhance the efficiency of intelligent manufacturing by 20%, cut the cost down by 20%, and save energy/reduce emission by 10% [5].

In the field of agriculture, IoT cuts the personnel costs for crop cultivation by about 50% and improves the overall economic benefit by about 10%. High precision environmental control in greenhouse facilities can be realised with the help of sensor-based automatic adjustment, and the high-quality green vegetables products cultivated for a high-end customer segment are priced 10 times higher than normal green vegetables [6].

In the field of energy conservation and environment protection, dynamic energy efficiency models can be established based on large data through energy management virtualisation, which can precisely locate the peak and valley electricity consumptions and then balance the peak and valley consumptions to save energy and reduce emission. For large industrial parks, the lighting energy reduction alone can be reduced by more than 30% [7].

In areas of urban management, pipe network monitoring, and intelligent transportation, the IoT has greatly enriched the urban management instruments and enhanced the urban management capacity. In transportation, 65% of the buses and near 70,000 taxis [8], passenger cars and dangerous chemicals vehicles in Beijing have been fitted with satellite positioning equipment, and five taxi monitoring centres and rail traffic control centres have been set up for intelligent management of all kinds of transportation.

10.3.3 IoT Trends and Standards

China has acquired important knowledge in network architecture, new types of sensors, M2M and other technologies. The WIA-PA (Wireless Networks for Industrial Automation – Process Automation) has been applied on a large scale in the oil and the electricity areas. The Huawei LTE-M system, which features low power consumption, low cost, low data rate and wide coverage, meets the needs of M2M applications and is now in the experimental stage for business deployment.

In the area of network structure, the release of the international standard ITU-T Y.2068 led by CAICT was completed in 2015. The Wuxi IoT Industry Research Institute and the China Electronics Standardisation Institute under the MIIT jointly promoted the approval of the ISO/IEC 30141 project, and have also proposed a consistent system decomposition model and an open standard design framework.

In the area of MEMS (Micro-Electro-Mechanical Systems) sensors, China's sensor enterprises sensibly grasped the new needs and new technologies of MEMS sensors, and have developed core technologies such as the MEMS acceleration meter technology, the MEMS sensor chip development and production test technology based on proprietary thermal detection method, the 5 million pixel CMOS image sensor based on the back lighting technology, the CMOS-MEMS process and the wafer level integrated package process. The first pilot-scale production line for the complete MEMS process has been built and the manufactured systems have been widely used in security monitoring, automotive electronics, consumer electronics and other fields.

For M2M (Machine-to-Machine) network platforms, both China Mobile and China Telecom are vigorously promoting the construction of M2M platforms. Studies on the optimisation of the existing networks and the M2M narrowband networks represent the current focus of activity. China will continue to promote standardisation work for network optimisation, including terminal triggering, low power consumption and wide coverage, as well as network congestion. Huawei and other device manufacturers have been carrying out research and development on narrowband M2M business supporting technologies, and have promoted the standardisation of the narrowband network.

10.3.4 The Internet and the Reconstruction of the Industrial Ecology

Chinese enterprises have demonstrated strong innovation ability in application services and business models. With the mobile Internet extending to IoT in recent years, Chinese Internet enterprises have emerged as the most dynamic powers in the development of IoT, and have been strongly influencing the patterns, models and industrial ecological system of China's IoT development. Major Chinese Internet companies have entered the field of IoT through wearable intelligent terminals, smart home, mobile health care, IoV, security, and other businesses, and have made rapid development in some of these areas.

On one hand, IoT applications can expand to be national-level applications in no time by virtue of mobile Internet portals and the large user scale. On the other hand, mobile APPs have become the data aggregation centres and feedback nodes for IoT. The anti-lost devices for children are integrated with a Bluetooth function, indicating children's distance from their parents, an alarm function when children are being beyond safe distance of their parents, and a four-fold location function, allowing parents to know the locations of their children at any time from a mobile APP.

10.3.5 EU-China Cooperation Proposal in IoT

The EU-China joint white-paper, published in January 2016 [3], has provided a list of cooperation items. The main ones are presented next.

10.3.5.1 Policy Level Cooperation

Encourage and actively promote research and innovation cooperation, and publication of results. Improve the EU-China cooperation policy and mechanisms in scientific research and innovation from a strategic and operational perspective, for elaborating policy recommendations. Encourage enterprises, institutions, and individuals on both sides to actively participate in cooperation projects and to form a long-term cooperation mechanism between the EU and China. At a later stage, and given that conditions are right in terms of fully reciprocal access to each other's RDI programmes, joint undertakings and calls will be considered as a further step.

The mechanism should be installed on two levels: governmental level and project level, preferably on a larger scale. For the first mechanism, policies should be investigated on both sides and provide input for the yearly EU-China ICT Dialogue. For the second mechanism, a wider scope of beneficiaries shall be considered including IoT Large Scale Pilots and Megaprojects.

10.3.5.2 Technical Cooperation

Carry out twinning activities between IoT Large Scale Pilots and Mega projects on IoT key technologies such as the IoT architectures, test-beds and platforms, semantic and technical interoperability, thus making full use of the knowledgebase and advantages of both regions. Encourage enterprises to carry out technical cooperation in strategic sectors on key product development, which can help each of the parties involved to break through technical bottlenecks and promote the process of high-tech industrialisation on a reciprocal basis. Expertise can be enhanced and cultivated through short, medium and long-term exchanges of PhD and post doctorial students, faculty staff, industry researchers. This should also be considered for entire institutes and companies.

10.3.5.3 Standards Cooperation

Encourage EU-China mutual support and jointly push the development of international standards for the IoT business layer, in the activities of international standardisation organisations such as OneM2M, ETSI, CEN/ ISO, IEEE, ITEF and ITU-T. A joint position paper on EU-China IoT standardisation mapping including recommendations should be elaborated, which can thus provide a reference for the future EU-China standards cooperation. This should also include a consideration of domain specific standards which could be done in conjunction with large scale projects as mentioned in previous section.

10.3.5.4 Market Cooperation

Strengthen EU-China information exchange and cooperation between the technology innovation strategic alliance of the IoT industry in China and Alliance for IoT Innovation in the EU to establish an effective market supply and demand platform for European and Chinese enterprises, which can expand bilateral industrial research and innovation activities. Joint market analyses of potential applications of IoT in diverse fields are needed to instate confidence. Mutual studies on topics related to IoT large scale projects could be a means of providing this confidence.

10.4 Adapting IoT to New Needs: Challenges from Brazil

The IoT has received a lot of attention in Brazil in the last years by the academic communities, companies and Brazilian agencies. Innovation and entrepreneurship communities formed by start-ups and SMEs are very active in IoT. In fact, IoT has been viewed as one of the best opportunities in decades to foster Brazilian economic and social development. In order to promote IoT development, Brazilian funding institutions have also provided funding lines addressing IoT, Smart Cities, Smart Utilities and Advanced Manufacturing. Large companies such as Intel, Huawei, Cisco, Ericsson, IBM and Samsung, have invested in RD&I centres in Brazil. Joint calls EU-Brazil are also an important mechanism to foster research in IoT and IoT related technologies and applications.

However, the adoption of IoT in Brazil has been slow, mainly due to the required investments. In order to foster IoT development, it is still necessary to address also points like human resources education and training, standardisation, interoperability, adoption of open standards and platforms, privacy and security, among others. Furthermore, Brazil needs to increase its participation in international standardisation bodies and also get better conditions to collaborate with international initiatives.

10.4.1 IoT RD&I Funding in Brazil

Recent calls for proposals from important funding institutions, such as FAPESP (São Paulo Research Foundation), addressing Smart Cities innovative projects, and BNDES (Brazilian Development Bank), addressing a technical study to diagnose and propose a public policy to foster IoT development and application in Brazil, show the importance Brazilian agencies are starting to give to IoT.

The BNDES calls for proposals complements other efforts at Federal government level in IoT. Additional actions include: 1) elaboration by the Brazilian Agency for Industrial Development (ABDI) of a technical study diagnosing the Brazilian industrial competence in Smart Grids and Smart Cities; draft document of these studies have mapped the Brazilian ICT Industrial Supply-Chain Smart Grids [9]; and 2) Ministry of Communication (MiniCom) that coordinates the special chamber for M2M/IoT matters. The actions are not completed, however funding agencies such as FINEP (Funding Authority for Studies and Projects) and BNDES are members of the chamber and certainly will be looking for respond on the demands of the M2M/IoT chamber.

10.4.2 IoT Success Cases in Brazil

IoT applications in Brazil normally have been made in small scale, and the corresponding business models are to be better studied and developed. Many different types of development have been made, ranging from medical support, agriculture, smart cities and smart grids. Currently, in Brazil, IoT use cases have been more focused in logistics, applying Real Time Location Systems (RTLS) technology. Today there are many systems in development and/or in operation. One system (Clever Care) made by Kidopi start-up (http://kidopi.com.br/) in the medical area was considered by ONU (World

Summit Award – United Nations – World Summit on the Information Society 2015) one of the five better medical applications in the world.

10.4.2.1 RFID/IoT Change of Paradigm

The most successful case of IoT application in Brazil corresponds to BRAS-COL, a wholesale company. BRASCOL successfully employed RFID to introduce a new management of inventory model. It reached a tremendous success in term of performance and cost savings. And this was possible because the owners took the risk to identify all the products that they sell independent of its sale price. This decision was made against the "orthodox thinking" that each product should be able to pay for its cost. The gains in the total operation were very significant proving that it was a savvy decision.

10.4.2.2 Smart Metering and Smart Grids

As a result of the effort that has been made by the government and companies in order to achieve a better energy efficiency. It is expected a roll-out of smart-meters in the electric sector, starting in 2016 with 1 million units in the state of Rio de Janeiro, going to 2 million by 2017 in the state of São Paulo and reaching its peak by 2020. Notably, the Total Available Market (TAM) of Brazil in respect to smart meters is of 78 million units.

In Smart Grids, the ABNT (Brazilian Association of Technical Standards) along with the COBEI (Brazilian Committee of Electronic, Electricity and Lighting) are working on standards for Smart Meters through 8 working groups. In telecom, several companies are conducting studies and trials on LPWA (SigFox by Telefonica Vivo Brazil) and LoRa (Unitec).

10.4.3 International Standardisation Related to IoT

In the international scenario, it is observed a large number of IoT related standardisation initiatives: ISO, OMG, ITU, IEEE, etc. These standardisation efforts compete with each other in standards. Being a founding member of ISO the Brazilian National Standards Organisation (ABNT) follows ISO Standards. Historically, Brazil have not been a strong participant in standardisation entities. In the IoT, there is some activity only in specific application areas, such as Smart Grids and Telecommunications.

10.4.4 EU-Brazil Collaboration on IoT

EU-Brazil cooperation in the area of ICT is regarded as having a crucial strategic value and high societal impact. It has been developing since the launch of the first coordinated joint call back in 2011. The cooperation is supported by an EU-Brazil dialogue on Information Society with specific working groups in some areas addressing not only research and innovation matters but also ICT policy and regulatory aspects. The main activities in the IoT are presented next.

10.4.4.1 EU-Brazil Joint Call for IoT Pilots RIAs

The Brazilian government and the European Commission have decided to launch the 4th coordinated call, to open in November 2016, in two main areas: cloud computing and IoT Pilots. Priorities for the call and for the development of future technologies are the 3O's: Open data, Open platform, Open science.

The IoT pilots call is to fund actions that validate and demonstrate IoT approaches and already developed IoT technologies and tools, to specific socio-economic challenges in real-life settings. The call will support three projects proposing pilots in five areas of interest, namely: (1) environmental monitoring; (2) smart water management; (3) energy management at home and in buildings; (4) smart assisted living and wellbeing; and (5) smart manufacturing focused on customisation.

Pilots are expected to empower citizens, both in the public and private spheres, and businesses, as well as improve the associated public services, for improved sharing of information, approaches and solutions, as well as expertise. Pilots should take place on both sides and across the Atlantic, involving end-users, establish common benchmarks, contribute to standardisation, open-source and open-data repositories and link with ongoing work in the IoT Focus Area.

10.4.4.2 EU-Brazil Mapping and Comparative Study

European stakeholders are supporting Brazil in the context of the sectorial dialogues for cybernetic policy on the development of the M2M/IoT ecosystem by performing an EU-Brazil mapping and comparative study. The action is being promoted by the Ministry of Communications of Brazil.

The general objectives of the project are: – establish the basis for the participation of Brazil together with Europe in the development of policies and regulation to overcome any trade, technological or legal barriers that might hamper the development of the IoT ecosystem; – collaborate with setting IoT standards and features in Brazil and Europe; – Extend the existing collaborative research between Brazil and Europe; – Brazil's participation and contribution to future cooperation agreements for research; – Harmonize actions between the Brazilian and European IoT chambers.

The outcomes are expected to: (i) provide valuable information for the development of Brazilian public policies for the promotion and application of the IoT/M2M ecosystem. (ii) improve capacity of the Brazilian state for international cooperation and joint action in the field of telecommunications and IoT platform applications; (iii) be input to define concrete steps to integrate an action plan (roadmap) of collaboration; (iv) be a contribution to define possible agreements for research activities and joint work between the thematic chambers of M2M/IoT from Brazil and Europe.

10.4.4.3 The EU-Brazil FUTEBOL Project

The H2020-688941 FUTEBOL project works towards the creation of a federated control framework to integrate test beds from Europe and Brazil for network researchers from academia/industry with unprecedented features. The major goal is to allow the access to advanced experimental facilities in Europe and Brazil for research and education across the wireless and optical domains.

The FUTEBOL project consortium argues that the needs of future telecommunication systems, be it from high data rate applications in smart mobile devices, machine-type communications (M2M) and the IoT, or backhaul requirements brought about from cell densification, require the co-design of the wireless access and the optical backhaul and backbone.

As an example, FUTEBOL will integrate the Bristol-is-Open (BiO) city-scale and real-life test bed and will offer it to experimenters. BiO supports IoT and data centre infrastructure integration with the wireless and optical backbone of a city infrastructure ecosystem. This will create opportunities for wider adoption of FUTEBOL's experimental facility, both within smart cities and the wider industry.

10.4.4.4 Further Work on EU-Brazil Cooperation

Since Horizon 2020, Brazilian entities are not entitled to receive funds from the European Union. This situation weakened the presence of SME's as the Brazilian structure of funding does not allow any type of company to get funds from the federal government. So, the EU needs to keep working with Brazil to find other ways for this funding. One option is to work more synchronised with FAPESP and Foundations from other states also.

Also it would be very useful to get joint works that can map the Brazilian companies that are interested in cooperative projects and their interest areas. The Brazilian IoT Forum has done some works in this direction and is now in a process to assemble an International Advisory Committee and could be an agent to disseminate in Brazil this collaboration EU-Brazil in IoT.

10.5 Do More with Less: Challenges for Africa. Low-Cost IoT for Sub-Saharan African Applications

ICT in the African context must be seen as a horizontal enabler in all areas of service delivery: eHealth, eGovernment, eAgriculture, eEnvironment, eEducation and eInfrastructures [10]. In several cases, ICT has enabled convergence of productive sectors, serving as platform for more holistic development. In fact, there are many examples of ICT developments in Africa that cut across traditional sectors: notable examples are the introduction of micro-health insurance and health-savings accounts through mobile devices; index-based crop insurance; crowd-sourcing to monitor and manage the delivery of public services. These innovative applications – for several reasons more disruptive in social terms than many counterparts in the EU – recognize and leverage commonalities between sectors, blur traditional lines, and open up a new field of opportunities.

Most of ICT success stories in Africa address very concrete issues of local populations. For instance, it is reported that 70% of the population of Senegal relies on cattle raising as their main source of revenues. When those animals are stolen, some families are left in such dramatic situation that cases of suicide are not unheard of [11]. DARAL [12] was a first attempt to fight against cattle rustling with the help of technology. For instance, it provides a web application for cattle identification and is currently implemented in 5 zones with 1500 farmers and 18000 cattle registered. DARAL emerged from an initiative of Coders4africa where 5 teams of 4 developers worked from collecting end-user requirements to the development of the final application. DARAL was one of those. For now, the current system is mostly a human-based cooperative alerting system based on SMS exchanges but automatisation can be foreseen by integrating active communicating components in the process, following the IoT trends.

Therefore, the opportunity of IoT applications in Africa is huge and it is not a question any more on whether IoT will come or not: many companies have already defined internal business activities to go along with this global move. However, when developed countries discuss about massive deployment of IoT, African countries are still far from being ready to enjoy the smallest benefit of IoT: lack of infrastructure, high cost of hardware, complexity in deployment, lack of technological eco-system and background, etc. [13].

In Sub-Saharan Africa about 64% of the population is living outside cities. The region will be predominantly rural for at least another generation. The pace of urbanisation here is slower compared to other continents, and the rural population is expected to grow until 2045. The majority of rural residents manage on less than few Euros per day. Rural development is particularly imperative where half of the rural people are depending on the agriculture/micro and small farm business, other half faces rare formal employment and pervasive unemployment. For rural development, technologies have to support several key application sectors like health, water quality, agriculture, livestock farming, climate change, etc. Therefore, when deploying IoT in Sub-Saharan African countries, it is necessary to target the removal of three major barriers: (1) Lower-cost, longer-range communications; (2) Cost of hardware and services; and (3) Limit dependency to proprietary infrastructures, provide local interaction models. These are further detailed next.

10.5.1 Lower-Cost, Longer-Range IoT Communications

Vast distances and poor infrastructure isolate rural areas, leaving those who live there poorly integrated into modern ICT ecosystems. Deploying IoT in this context must use longer range wireless communication to decrease both the complexity and the cost of data collection. Using the telco mobile communication infrastructure, when coverage is available, is still very expensive (e.g., GSM/GPRS) and definitely not energy efficient for autonomous devices that must run on battery for months. Recent so-called Low-Power Wide Area Networks (LPWAN) such as those based on Sigfox$^{(TM)}$ or Semtech's LoRa$^{(TM)}$ technology definitely provide a better connectivity answer for IoT as several kilometres can be achieved without relay nodes to reach a central gateway or base station. When adding the financial cost constraint and the network availability, LoRa technology, which can be privately deployed in a given area without any service subscription, has a clear advantage over Sigfox which coverage is entirely operator-managed. Some LoRa community-based initiatives such as the one promoted by The Things Network [14] may provide interesting solutions and feedbacks for dense environments such as cities but under the agriculture/micro and small farm business model an even more ad-hoc and autonomous solution need to be investigated and deployed. On the software side, the software service platform will also need to offer highly innovative monitoring, recommendation, notification services based on the data coming from multiple rural application sectors, taking into account that, in most cases, the mobile phone is the unique technological terminal for end-users.

10.5.2 Cost of IoT Hardware and Services

The maturation of the IoT market is happening in many developed countries: innovative and integrated products are available for smarter home and various monitoring applications. While the cost of such devices can appear reasonable within developed countries standards, they are definitely still too expensive for very low-income sub-Saharan ones. The cost argument, along with the statement that too integrated components are difficult to repair and/or replace definitely push for a Do-It-Yourself (DIY) and "off-the-shelves" design orientation. To be sustainable and able to reach previously mentioned rural environments, IoT initiatives in developing countries have rely on an innovative and local business models. We envision mostly medium-size companies building their own "integrated" version of IoT for micro-small scale services. In this context, it is important to have dedicated efforts to design a viable exploitation model which may lead to the creation of small-scale innovative service companies.

The availability of low-cost, open-source hardware platforms such as Arduino-like boards is clearly an opportunity for building low-cost IoT devices from mass-market components. For instance, boards like Arduino Pro Mini based on an ATmega328 microcontroller offers an excellent price/performance/consumption trade-off and can be used to provide a low-cost platform for generic sensing IoT with LoRa long-range transmission capability. In addition to the cost argument (cost can be less than 15 euro for a fully operational long-range sensing device) such mass-market component greatly benefits from the support of a world-wide and active community of developers. See in Figure 10.1 the experimental set-ups with Arduino Pro Mini.

With the gateway-centric mode of LPWAN technology, commercial gateways are usually able to listen on several channels and radio parameters

Figure 10.1 Generic platform with Pro Mini (left), packaged for battery-operated and outdoor deployment (right).

simultaneously. They use advanced concentrators radio chips that alone cost more than a hundred euro. Here, again, the approach can be different in the context of agriculture/micro and small farm business: simpler "single-connection" gateways can be built based on a simpler radio module, much like an end-device would be. Then, using Linux-based platforms such as the Raspberry PI that has high price/quality/reliability trade-off, the cost of such gateway can be less than 45 Euro. See in Figure 10.2 the prototypes of the low-cost gateway.

Therefore, rather than providing large-scale deployment support, IoT platforms in developing countries need to focus on easy integration of low-cost "off-the-shelves" components with simple, open programming libraries and templates for easy appropriation and customisation by third-parties. By taking an ad-hoc approach, complex mechanisms, such as advanced radio channel access to overcome the limitations of the low-cost gateway, can even be integrated as long as they remain transparent to the final developers.

10.5.3 Limit Dependency to Proprietary Infrastructures, Provide Local Interaction Models

Once data are collected on the gateway, they usually have to be pushed/uploaded to some Internet/cloud servers for storage and visualisation; and eventually for further processing tasks. It is important in the context of developing countries to be able to use a wide range of infrastructures and, if possible, at the lowest cost. Fortunately, along with the global IoT uptake, there is also a tremendous availability of sophisticated and public IoT clouds platforms and tools [15], offering an unprecedented level of diversity

Figure 10.2 Several versions of the low-cost gateway (left), close-up view on the PoE version for easy integration into existing network infrastructures (right).

which contributes to limit dependency to proprietary infrastructures. Many of these platforms offer free accounts with limited features but that can already satisfy the needs of most agriculture/micro and small farm business models we are referring to when addressing IoT for Sub-Saharan African applications. What are the impacts on the design architecture/choices of the deployed IoT platforms? One simple design orientation is to highly decouple the low-level gateway functionalities from the high-level data post-processing features, privileging high-level languages for the latter stage (e.g., Python) so that customizing data management tasks can be done in a few minutes, using standard tools, simple REST API interface and available public clouds.

One additional important issue that needs to be taken into account in the context of sub-Saharan Africa is the lack or intermittent access to the Internet. Data should also be locally stored on the gateway which can be directly used as an end computer by just attaching a keyboard and a display. This solution perfectly suits low-income countries where many parts can be found in second markets. The gateway should also be able to interact with the end-users' smartphone through WiFi or Bluetooth to display captured data and notify users of important events without the need of Internet access as this situation can clearly happen in very remote areas.

10.5.4 The H2020 WAZIUP Project

Most of the challenges illustrated in here are planned to be addressed in the H2020-687607 WAZIUP project. The WAZIUP project, namely the Open Innovation Platform for IoT-Big Data in Sub-Saharan Africa is a collaborative research project using cutting edge technology applying IoT and Big Data to improve the working conditions in the rural ecosystem of Sub-Saharan Africa. First, WAZIUP operates by involving farmers and breeders in order to define the platform specifications in focused validation cases. Second, while tackling challenges which are specific to the rural ecosystem, it also engages the flourishing ICT ecosystem in those countries by fostering new tools and good practices, entrepreneurship and start-ups. Aimed at boosting the ICT sector, WAZIUP proposes solutions for long term sustainability. See Figure 10.3 for the project's general technical approach built around the low-cost gateway and (Future) Internet technologies.

The consortium of WAZIUP involves 7 partners from 4 African countries and partners from 5 EU countries combining business developers, technology experts and local Africa companies operating in agriculture and ICT. Central to WAZIUP's concerns is the inclusion of developer communities

Figure 10.3 The gateway can push data to Internet cloud resources (left) or provide local connectivity with WiFi-based web server or Bluetooth-based smartphone app (right).

(e.g., Coders4Africa) and innovation hubs (e.g., CTIC, iSpace) who have experience to train, adapt, validate and disseminate results. Quick appropriation and easy customisation by third-parties is ensured by tightly involving end-users' communities in the loop, namely rural African communities of selected pilots, and by frequent training and hackathon sessions organised in the sub-Saharan African region.

10.6 EU-Japan Collaboration for a World Leading Research in IoT

The world is facing a number of critical challenges such as global warming, economic crisis, security threats, inequality, natural disasters and ageing society. Urban areas are particularly affected, given that the world population is increasingly concentrated in those areas. Currently more than 75% of the population in Europe and more than 90% of the population in Japan live in urban areas[2]. Further, those areas are expected to absorb the majority of the population growth expected over the next four decades, while at the same time drawing in some of the rural population, thus world population in urban areas is expected to be 66% by 2050.

While occupying 2% of the earth's surface, cities use 75% of the world resources. Those resources in civil infrastructure such as water, energy, public transportation, parking spaces, buildings, roads, bridges, etc., as well as natural resources and economic resources need to be shared by this increasing population. This has direct consequences for urban citizens and for the city itself.

Ranging from social to economic aspects, IoT provides countless possibilities to enhance the quality of life and security of people, while at the same time reducing inequalities and providing new revenue opportunities for enterprising businesses, from large groups and public administrations to SMEs, start-ups and web entrepreneurs. Considering this potential, European Commission and two Japanese funding agencies, namely NICT (National Institute of Information and Communication Technologies) and MIC (Ministry of Internal Affairs and Communication) have launched the first joint call for projects on IoT in 2012 in the context of the FP7 Programme. It is followed by two other calls in 2014 and in 2015 within the H2020 programme. The following sections give an overall summary of three projects from those calls, namely ClouT, FESTIVAL and iKaaS.

---

[2]Uexküll, Jakob. Shaping our future: Creating the World Future Council. Foxhole, Devon.

10.6.1 ClouT: Cloud of Things for Empowering Citizen ClouT in Smart Cities

ClouT is a collaborative Europe – Japan project that has developed a smart city platform which benefits from the latest advances in IoT and Cloud Computing domains. ClouT, which stands for Cloud of Things, provides a virtualisation framework to provide a uniform way of representing various city data sources such as IoT devices, legacy devices, social networks, mobile applications and World Wide Web. Based on a reference Cloud + IoT architecture and smart city domain model, ClouT platform has been developed allowing secure access to real-time data as well as historical data with easy-to use tools targeting municipalities, citizens, service developers and application integrators to create, deploy and manage smart city applications. The ClouT project has been jointly coordinated by CEA and NTT East, and it is further bringing together prestigious private companies such as ST Microelectronics, Engineering IngegneriaInformaticaSpA, Panasonic, NTT R&D as well as academic institutes such as University of Cantabria, Keio University and National Institute of Informatics, which have strongly committed to bring this first EU-Japan initiative on IoT and cloud to a success.

The project has developed several smart city applications using the developed platform and tools and deployed them in 4 pilot cities of the project: Santander, Genova, Fujisawa and Mitaka. Applications e.g., environmental monitoring, context aware coupons, city dashboards, citizen safety applications, elderly care social networks, have been validated via field trials involving real end-users. See in Table 10.2 a few examples.

The project has provided its outcomes in terms of deliverables, reusable software, fried trial descriptions, newsletters, videos, etc. All the information from the project is available at the project website: http://clout-project.eu.

10.6.2 FESTIVAL Federated Interoperable Smart ICT Services Development and Testing Platform

There have been long years of research work in Europe and Japan on federation of test beds and more recently on IoT test beds. FESTIVAL aims at leveraging those test beds by a federation approach where experimenters can seamlessly perform their experiments taking benefit of various software and hardware enablers provided both in Europe and in Japan. Facilitating the access to those test-beds to a large community of experimenters is a key

Table 10.2   Field trials performed in the ClouT project

Fujisawa: Sensorised garbage cars

This application aims to collect atmospheric information by mobile sensor system installed on garbage collection cars in Fujisawa City. Fujisawa municipality can monitor the location and operational status of each garbage collection car through the Control Center application.

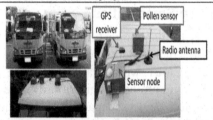

Santander: Smiley Coupon

After the successful trial of the Smiley Coupon in Fujisawa, ClouT replicates it in the Santander city. The application provides customized coupons for the citizens and visitors according to their degree of smile. Commercial firms (restaurants, bars, shops, etc.) participate with a wide range of special offers.

Mitaka: Sanpoki Stamp rally

The application contributes to the prevention of isolation of young and elderly people, by encouraging them to go out and walk through suggested routes that match their interests. It allows to post attractive photos about Mitaka and share information among the citizens.

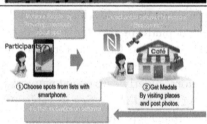

Genova: "I don't risk" application

This application informs citizens about good practices and general information about environmental risks and emergency situations by using environmental data from weather sensors, hydrometers, webcams, etc. It has become one of the top mobile applications of the Genova City with more than 4000 downloads and average rating of about 4/5.

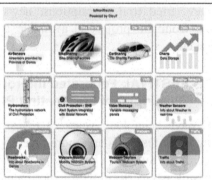

asset to the development of a large and active community of application developers, necessary to address many challenges faced by European and Japanese societies.

FESTIVAL is a H2020 European-Japanese collaborative project that aims to federate heterogeneous IoT test beds, making them interoperable and building an "Experimentation as a Service" (EaaS) model. FESTIVAL test beds connect cyber world to the physical world, from large scale deployments at a city scale, to small platforms in lab environments and dedicated physical spaces simulating real-life settings. IoT is related to the physical world, thus real-life conditions are essential to validate the IoT applications. The involvement of end-users is also of tremendous importance to validate the quality of user experience. Going beyond the traditional nature of experimental facilities, related to computational and networking large scale infrastructures, FESTIVAL test beds have heterogeneous nature and in order to be federated they have been clustered in four categories: "Open Data" (i.e., open datasets), "IoT" (i.e., sensors and actuators), "IT" (i.e., computational resources) and "Living Labs" (i.e., people). Figure 10.4 illustrates the FESTIVAL's federation architecture.

Considering that every test bed category provides specific resources, the main challenge for FESTIVAL is to develop a platform that can allow experimenters to access very different assets in a homogeneous and transparent way, supporting them in the phases of the experiments. The architecture aims at providing the blueprint to be used to build the federated FESTIVAL test bed. It specifies a common resource data model and a set of uniform APIs that will be used by the experimenters to build and deploy rapidly and efficiently their experiments. Thanks to the FESTIVAL's uniformed approach, the experiments will be portable across several test beds and replicable with minimum effort of adaptation.

Furthermore, FESTIVAL tools include the possibility to access FIWARE Generic Enablers allowing to deploy predefined components to address specific needs in the experimentation (e.g., data analysis, big data management etc.). The FESTIVAL platform will be tested on various application domains across Japan and Europe such as smart city, smart energy, smart building and smart shopping.

FESTIVAL project is jointly coordinated by CEA and Osaka University and brings together 12 other institutions from Europe and Japan (www.festival-project.eu).

Figure 10.4   The overall view of the FESTIVAL architecture.

10.6.3 iKaaS – Intelligent Knowledge as a Service

Smart Cities are happening, and while the increasing sensor deployment in urban public and private spaces provides invaluable data about resources and services demand, the sheer amount of data that is available in data bases and data stores or that can be collected through IoT influences urban life is staggering. Understanding this data, deriving knowledge from it to improve service provision as well as usage of resources is of the utmost importance. And moreover, the lessons that can be learned in one city and the knowledge derived can be applied to other cities, in other parts of the world.

However, independent of the location, this relies on the participation of the citizens and they will only be willing to provide personal information if their own data is secure and is being kept private, both before and after knowledge has been derived from it. The iKaaS (intelligent Knowledge-as-a-Service) project brings the essential building blocks for this together; it defines a platform that integrates the three concepts of cloud computing – big data analytics – IoT. iKaaS defines and builds a secure data storage and privacy-preserving analytics engine over heterogeneous multi-cloud environments spanning across national borders.

As user participation, and personal data sensed around and about the user, are at the core of building and operating such a knowledgebase, the iKaaS platform builds privacy, security and trust into storage, access and analysis capabilities already "by-design" rather than as an add-on. It implements technical and organisational measures and procedures in such a way that the processing will ensure the protection of the rights of the user (citizen). And this also includes the definition and implementation of mechanisms that help ensuring that privacy is preserved even when personal data have been processed. The iKaaS approach applies privacy-preserving data release methods that guarantee some anonymisation, the iKaaS approach goes beyond this and builds empirical models to quantify the risks associated with those methods, and relates those risks using the notion of "costs of attacks".

iKaaS brings together cloud computing – big data analytics – IoT to derive knowledge, the project intends to apply these means to platform instances in different cities, across different countries and across the boundaries of administrations and data regulation. These form clear implementation challenges, especially over multiple cloud environments in different administrative domains and a myriad of connected personal computing devices. The iKaaS platform will cater for applications built atop a knowledgebase to provide end-user as well as business-to-business or business-to-government services.

iKaaS tackles three use cases around the wider topic of personal and public health, as this implies that citizens' personal health related data is being used to derive new knowledge, the requirements to data and privacy protection are immense. However, at the same time, the knowledge that can be gained from personal data together with environmental observations (air quality, weather conditions, etc.) will help the wider community to improve conditions or prevent individual exposure to potentially harmful conditions/situations. Based on the existing regulations for the treatment of personal information among member countries, iKaaS investigates solutions for flexible and privacy enhanced treatment of cross border data which is transitioned via iKaaS platform. This includes demarcation points of responsibility of data holders, data transfers or data receiver's and remedies if problems occur. Via a multi stake-holder scheme, iKaaS defines best practices for privacy and data protection treatment of cross border data.

Multi-Cloud Architecture: iKaaS designs an open, adaptable and secure Everything-as-a-Service framework for incorporating optimal service deployment which includes migration and parallelisation as well as distributed management of smart objects, associated storage, processing and communication of data, targeted to enable re-usability of applications across different domains and platforms as well as Knowledge-as-a-Service.

Knowledge as a Service: iKaaS investigates and develops mechanisms that facilitate re-use of smart objects as a distributed data processing capability, across different administrative and business domains. iKaaS also develops mechanisms to analyse data and derive Knowledge-as-a-Service (KaaS).

Security, Privacy, and Trust: iKaaS designs an open, adaptable and secure Everything-as-a-Service framework for incorporating optimal service deployment which includes migration and parallelisation as well as distributed management of smart objects, associated storage, processing and communication of data, targeted to enable re-usability of applications across different domains and platforms as well as Knowledge as a Service.

To reach its aims and implement the iKaaS platform, the project team requires expertise and partners from various domains. The consortium is coordinated by the University of Surrey and KDDI R&D Labs and consists, altogether, of six partners from Japan and nine partners from European Countries, their skillset and expertise are complementarity in the specified iKaaS problem domains. iKaaS demonstrates their use cases in Sendai, Japan (i.e., the town of Tago-Nishi) as well as in Madrid, Spain. (www.ikaas.com).

10.7 EU-US IoT Cooperation

Today there are two main initiatives for the IoT created at global level and organised as alliances/consortia: the Industrial Internet Consortium (IIC) and the Alliance for the IoT Innovation (AIOTI). Both these IoT alliances/consortia create unique value with their organizational entities, by engagement, by stimulating and matchmaking relationships between companies creating new applications, increasing revenue, industry reach and shared knowledge and experience, and support for a long-term value-creating, collaborative relationship, leading to success for the partners involved as well as for the eco-system as a whole.

IIC, founded by AT&T, Cisco, GE, IBM, and Intel, brings together the organizations and technologies necessary to accelerate the growth of the Industrial Internet by identifying, assembling and promoting best practices. Membership includes small and large technology innovators, vertical market leaders, researchers, universities and government organizations. The goals of IIC are to:

- Drive innovation through the creation of new industry use cases and test beds for real-world applications
- Define and develop the reference architecture and frameworks necessary for interoperability
- Influence the global development of standards process for internet and industrial systems
- Facilitate open forums to share and exchange real-world ideas, practices, lessons, insights
- Build confidence around new and innovative approaches to security

The Industrial Internet Consortium Working Groups coordinate and establish the priorities and enabling technologies of the Industrial Internet in order to accelerate market adoption and drive down the barriers to entry. There are currently 19 Working Groups and teams, broken into 7 broad areas:

- Business Strategy and Solution Lifecycle
- Legal
- Marketing
- Membership
- Security
- Technology
- Test beds

These groups are comprised of Industrial Internet Consortium member company representatives. Member companies can assign an unlimited number of individuals to the Working Groups, which follows the one vote, one company rule.

The Alliance for IoT Innovation, AIOTI, was initiated following the European and global IoT technology and market developments and aims to create and master sustainable innovative European IoT ecosystems in the global context to address the challenges of IoT technology and *applications* deployment including standardisation, interoperability and policy issues, in order to accelerate sustainable economic development and growth in the new emerging European and global digital markets.

The AIOTI mission statement covers the following points:

- Develop IoT ecosystems across vertical silos including start-ups and SMEs
- Identify, communicate and champion EU spearheads to speed up the take up of IoT
- Mapping and bridging global, EU and Members States' IoT innovation activities
- Gather evidence on market obstacles for IoT deployment in a Digital Single Market context
- Contribute to Large Scale Pilots to foster experimentation, replication and deployment and to support convergence and interoperability of IoT standards.

AIOTI strategy translates the vision and mission into goals and actions that provide unique value by the Alliance to its stakeholders. Key strategic elements include:

- A unique *application*-driven IoT initiative bringing together the demand and supply side stakeholders beyond technology and complemented by horizontal research, innovation, standardisation and policy cross-cutting working structures
- A goal oriented Alliance aiming to be agile, flexible, lean and project driven applying clear stimulus measures among its members
- The European reference platform addressing IoT in the global context
- AIOTI aims to be strongly and firmly positioned in the global IoT landscape.

AIOTI Working Groups coordinate and establish the research, innovation priorities and enabling technologies in the area of IoT (consumer/business/industrial) in order to accelerate sustainable economic development and growth based on IoT technology and applications deployment and adoption. There are currently 13 Working Groups, broken into 4 horizontal groups and 9 vertical groups:

- IoT European Research Cluster (IERC)
- Innovation Ecosystems
- IoT Standardisation
- Policy issues
- Smart living environment for ageing well
- Smart farming and food security
- Wearables
- Smart cities
- Smart mobility
- Smart environment (smart water management)
- Smart manufacturing
- Smart energy
- Smart Buildings and Architecture

In this context, the cooperation between EU and US is very important. The mechanism of cooperation are installed on two levels: governmental level and project level, preferably on a larger scale. Policies should be investigated on both sides and provide input for the yearly EU-US ICT Dialogue. In addition, a wider scope of beneficiaries shall be considered including IoT Large Scale Pilots and IoT Test beds.

In this context there are a number of European projects and initiatives [18–21] that are addressing the EU-US cooperation and collaboration. PICASSO project created the framework to bring together experts in the field of 5G, Big Data, IoT, CPS to focus on identifying the key issues in each specific field and on policy issues that touch upon all of these domains. The ICT Policy Expert group will focus on Privacy and Data Protection, in recognition that policy issues relating to this touch all ICT developments across the Atlantic. UNIFY-IoT project as part of the European Platforms Initiative (IoT-EPI) is leading the task force on international cooperation in order to define the strategy and activities for international collaboration with global players working at initiatives and projects in the IoT domain. The task force is coordinating the activities to be planned and executed for liaising, interacting and then follow-up with the relevant projects' stakeholders and IoT ecosystems. The group is coordinating the interaction with international initiatives by supporting the IoT ecosystems to meet global challenges and to be adopted worldwide in order to be successful. The intent is to get a clear overview of the priority policy issues in ICT collaboration, and insights in how these issues can be addressed from a bilateral multi stakeholder perspective in a global context.

In the context of establishing liaisons with key stakeholders outside the EU the cooperation and meeting with the US stakeholders offer the possibility to present a panorama of the ICT and IoT landscape and programmes currently underway in Europe and the US as well as programmes in the rest of the world. Existing funding opportunities for collaboration are highlighted. The views of the EU-US Expert Groups on 5G Networks, Big Data, IoT, CPS are presented identifying gaps and opportunities, a map of challenges, open problems, and the needs for supporting policy measures and strategic EU-US initiatives (both policy and research related). Key actors, i.e., NIST, NSF, IMS, are involved together with the European projects and representatives from EC to highlight existing opportunities for collaboration.

10.7.1 Policy Level Cooperation

The IoT policy issues is addressed in Europe by 2014 European Commission's Article 29 Working Party on Data Protection [16] setting forth its interpretation of how EU data protection laws apply to IoT and in US by the 2015 Report on the IoT, from FTC [17] setting forth privacy and security best practices for IoT.

The WP 29 Report looks at IoT via EU data protection principles, highlighting these concerns for IoT manufacturers, developers and data collectors:

- Lack of control – Interconnectivity means a greater potential for automatic flow of data among devices (and vendors) without notice to users.
- Additional purposes – Interconnectivity also may lead to use of gathered data by third parties for other than the original intent.
- Consent – Because users lack full disclosure of data flow, their consent to initial data collection may be inadequate.
- Profiling – Fine-grained user monitoring and profiling could result from the type of information collectable from connected devices.
- Limiting anonymity – More use of connected devices suggests lower likelihood for maintaining anonymity.
- Security – Large volumes of data transferring over connected devices may lead to considerable security risks.

The WP 29 Report recommendations are the security and privacy concerns and recommends that IoT manufacturers, developers and data collectors:

- Conduct a privacy impact assessment before releasing a device.
- Delete raw data from the device as soon as it has been extracted.
- Follow privacy-by-design and privacy-by-default principles.
- In a user-friendly way, provide a privacy notice, and obtain consent or offer the right to refuse.
- Design devices to inform both users and people interacting with them (e.g., people being recorded by a camera in a wearable technology) of the data processing by the entity providing the device.
- Inform users of data that has been collected and enable them to access, review and edit that data before it is transferred.
- Give users granular choices on the type of processing as well as time and frequency of data gathering.

These principles apply whenever a connected device is used in the EU, even if the device did not originate in the EU. While the WP 29 Report is not binding law, it is persuasive to EU regulators, when deciding how to apply data protection law to the IoT. Once the new EU Data Protection Regulation takes effect, fines for violations of EU data protection law could be up to 5 percent of global turnover for a company. Thus, flouting the WP 29 Report principles, which are considered persuasive authority on the interpretation of EU data protection law, could result in very significant fines.

The FTC Report focuses on security (considered as harm to consumers from unauthorized access and misuse of personal information, attacks on other systems and safety risks) and privacy that are considered as following:

- Remote access to smart meters could enable thieves to determine when a house is empty, leaving it susceptible to robbery.
- A connected device could be used to gain control of a consumer's internal network and in turn, attack a third-party system.
- Remote access to stored financial data could enable fraud.
- Privacy-related concerns over the collection of sensitive information (geolocation, financial and health data), the sheer volume of data collected and the potential for misuse.

The FTC Report recommends best practices to IoT manufacturers, developers and data collectors, focusing on:

- Data security – The FTC recommends that device manufacturers adopt a privacy-by-design approach, including a privacy and security risk assessment made prior to release, use of "smart defaults" (e.g., forcing changes to default device passwords) and security and access control measures, and monitoring throughout the device's life cycle.

- Data minimization – While endorsing the necessity to limit collection and retention of users' data, the FTC calls for a "flexible approach," urging companies to "develop policies and practices that impose reasonable limits on the collection and retention of consumer data."
- Notice and choice – The FTC recognizes notice and choice play a "pivotal role," but – in contrast to the WP 29 view – acknowledges that notice and choice are not always necessary. Instead, the FTC calls for notice and choice where sensitive data is collected or where there is unexpected collection or sharing.

The EU-US expert groups, created by a number of European projects and initiatives [18–21], have identified different candidate policy issues as input for further bilateral discussions:

- Addressing global societal challenges, respecting Human Rights, supporting Sustainable Development Goals (SDGs),
- Trust and confidence, privacy and data protection encryption, censorship, surveillance, security, anonymity,
- Innovation ecosystem, start-ups, incubators, accompanying measures,
- (Open) standards, certification, transparency and choice.

These possible policy subjects are provided as a starting point, and are the baseline for the policy issues to be discussed in dialogue.

Trust and usability are very important success factors for IoT, and IoT security and privacy need to be addressed across all the IoT architectural layers and across the domain applications. Performance, complexity, costs are all factors which influence adoption in addition to those that engender trust. While there have been important progress made and actions planned to address usability there are nevertheless remaining a number of potential gaps in the overall "trust" framework that can be evaluated.

In this context the EU-US cooperation is seen at company level in the AIOTI Working Group 04 (WG04), where EU and US companies are addressing these issues. The AIOTI WG04, is to identify existing or potential market barriers that prevent the take-up of the IoT in the context of the Digital Single Market, as well as from an Internal Market perspective, with a particular focus on trust, security, liability, privacy and net neutrality.

10.7.2 Technical Cooperation

The IIC and AIOTI members could, in the future, maintain a technical exchange to identify mappings, research priorities, differences and enhancements, support the alignment of IoT architecture efforts for the benefit of interoperability of systems from the different domains, map of IoT reference architectures/platforms showing the direct relationships between elements of the models and a clear roadmap to ensure future interoperability.

AIOTI and IIC, as the global leading initiative frameworks for the IoT, create unique value with their organizational entities, by engagement, by stimulating and matchmaking relationships between companies. This approach is creating new applications, increasing revenue, industry reach and shared knowledge and experience, and support for a long-term value-creating, collaborative relationship, leading to success for the partners involved as well as for the IoT ecosystem as a whole.

Future EU-US cooperation can be seen in activities addressed in IoT Large Scale Pilots and IoT Test beds by discussing the main challenges related to IoT key technologies such as the IoT architectures, scalability and sustainability of large scale IoT deployments, IoT platforms, semantic and technical interoperability, thus making full use of the whole digital value chain and IoT applications and use cases deployed in both regions. In this context, the development of a common communication strategy that fully exploits the possible synergies between EU-US initiatives is important for the future collaboration.

10.7.3 Standards Cooperation

IoT is a global concept, and is based on the idea that anything can be connected at any time from any place to any network, by preserving the security, privacy and safety. The concept of connecting any object to the Internet could be one of the biggest standardization challenges and the success of the IoT is dependent on the use/development of interoperable global standards.

Global standards are needed to achieve economy of scale and interworking. Interconnected edge devices are evolving to intelligent devices, which need networking capabilities for a large number of applications and these technologies are "edge" drivers towards the IoT, while the network identifiable devices will have an impact on telecommunications networks.

Encourage EU-US mutual support and jointly push the development of international standards for the IoT business layer, in the activities of international standardisation organisations such as OneM2M, ETSI, CEN/ISO, IEEE, ITEF and ITU-T. Cooperation foreseen with NIST in the area of Smart Cities and application of IoT technologies in the cities. The cooperation could look at the development of performance standards, measurement tools, and guidance that enable city stakeholders and technology providers to design and implement effective solutions. The cooperation can address the coordination of the development of standards and guidelines for smart city interoperability and exchange experiences on smart city test beds or IoT large scale pilots.

The EU-US cooperation is coordinated at the company levels in the AIOTI Working Group 03 (WG03) that address IoT standardisation. The AIOTI WG03 has provided common views of the IoT stakeholders on the IoT standardisation that are covered in 3 documents "IoT Landscape and IoT LSP Standard Framework Concepts", "IoT High Level Architecture (HLA)", "IoT Semantic interoperability". The documents offer an extensive overview of the global IoT standardisation landscape allowing the stakeholders involved in IoTprojects to be flexible and innovative in their use of the information, while assuring that they provide standard-based and interoperable IoT implementations. The cooperation EU-US will extend on the alignment of requirements for standardization bodies to review and influence global standards.

10.7.4 Market Cooperation

Strengthen EU-US information exchange and cooperation between the technology innovation strategic alliance of the IoT industry in US like tie IIC and AIOTI in the EU to establish an effective market supply and demand platform for European and American companies, which can expand bilateral industrial research and innovation activities. Many European and American companies are members of both AIOTI and IIC. The EU-US cooperation at the level of alliances can support the exchange use cases and architectural requirements focused on industrial/business/consumer markets in order to meet the requirements in its specification for the different IoT solution implementations. The EU-US cooperation will focus as well on common support to accelerate the delivery of a cross sectorial IoT architectural framework (consumer/industrial/business).

10.8 Conclusions: Cooperation to Balance Globalisation and Differentiation of IoT Solutions Worldwide

Europe has devoted strong attention to international cooperation with the EU's partner countries and regions, developed on the basis of common interest and mutual benefit and create win-win situations. Many of these activities have been implemented to the appropriate scale and scope in the context of the Horizon 2020 framework programme. The IoT, in the large scope of ICT, has been further developing as a key area of international cooperation aiming especially at global IoT agreements but also on developing differentiation of IoT solutions for addressing specific needs and challenges for both EU and partner countries and regions.

The South Korean government has established the 'Mid- and Long-term R&D plan for IoT' that links existing R&D projects classified into units with the entire ecosystem. South Korea government strongly supports global collaboration with major countries including the EU. The WISE-IoT project has started, as part of jointly funded R&D programs, gathering lead contributors from Europe and South Korea to on-going major global IoT standardisation activities with the objective to strengthen and expand emerging IoT standards and reference implementation using feedback from user-centric and context-aware pilots. Further cooperation activities are expected in IoT standardisation and reference architectures but also on promoting the use of EU methodologies and models in the implementation of large-scale pilots in South Korea, especially in smart cities and healthcare application areas.

In China, the IoT has become an important carrier for strategic information industries and integrated innovation. The EU-China IoT Advisory Group, established in February 2011, is active on pushing global IoT standards while developing competitive IoT solutions. An EU-China joint white paper on IoT, published in January 2016, has laid down the areas of cooperation. Main ones include: (i) Policy level cooperation to encourage and actively promote research and innovation cooperation, and publication of results; (ii) Technical cooperation carrying out twinning activities between EU IoT Large Scale Pilots and China Megaprojects and enterprise-level cooperation in strategic sectors on key product development; (iii) Standards cooperation for EU-China mutual support and jointly push the development of international standards; and (iv) Market cooperation to strengthen EU-China information exchange and cooperation between the technology innovation strategic alliance of the IoT industry in China and Alliance for IoT Innovation in the EU.

EU-Brazil research cooperation in the area of ICT is regarded as having a crucial strategic value and high societal impact. It has been developing since the launch of the 1st coordinated call, back in 2011, to include a specific focus on IoT Pilots, in the context of the 4th coordinated joint call. Furthermore, Europe is supporting Brazil in the context of the sectorial dialogues for cybernetic policy on the development of the M2M/IoT ecosystem by performing an EU-Brazil mapping and comparative study. And, the EU-Brazil FUTEBOL project is working to create of a federated control framework to integrate test beds from Europe and Brazil to support network researchers from academia/industry looking out for the IoT and M2M future needs. The strategic cooperation of EU with Brazil is expected to be further supported and animated by the IoT Focus Area CSA project (to be awarded) on realisation of joint cooperation activities for active knowledge sharing and promotion of EU and Brazilian IoT ecosystems/technologies and the alignment of EU LSP and IoT pilots to be launched as part of the EU-Brazil research cooperation.

In respect to Africa, the opportunity for IoT applications is huge but African countries are still far from being ready to enjoy the smallest benefit of IoT due to the lack of infrastructure, high cost of hardware, complexity in deployment, lack of technological eco-system and background, etc. As such, and when deploying IoT in African countries, it is necessary to address three major barriers: (1) Lower-cost, longer-range communications; (2) Cost of hardware and services; and (3) Limit dependency to proprietary infrastructures, provide local interaction models. Most of these challenges are being practically addressed by the H2020-687607 WAZIUP project collaborative research project by using cutting edge technology applying IoT and Big Data to improve the working and living conditions in the rural ecosystem of Sub-Saharan Africa. International collaboration of EU with African countries, and particularly with South Africa, will pursue towards IoT approaches and solutions that especially address the development goals.

Europe and Japan are two leading economies which have the necessary potential to provide world leading technologies for smarter citizen life. The report "Digital Economy in Japan and the EU"[3] identifies common challenges between the European and the Japanese economies, including the "scaling up of smart city projects". To respond to those challenges and following the success of the preceding joint calls, European Commission and Japanese agencies have decided to continue collaboration in the context of the H2020 and launched new joint calls on, not only IoT and smart cities but also on related topics such as 5G, experimentation test beds, ICT-assisted well-ageing, cyber-security, etc. The first conclusions from the achieved projects confirm that Europe and Japan can provide a strong and reliable partnership to face together emerging social, economic and environmental challenges.

The EU-US cooperation will increase in the future in the area of IoT on several levels, governmental, alliances, companies and projects levels.

European Commission gives a strategic dimension to IoT for the Digital Single Market (DSM), not only in terms of regulatory challenges but also with regards to overcome interoperability issues and fragmented standards, probably one of the most dominant obstacles at the moment with the key objective to develop, implement and deploya collaborative, responsible and

---

[3] EU-Japan Centre for Industrial Collaboration, March 2015: http://www.eu-japan.eu/digital-economy-japan-and-eu-assessment-common-challenges-and-collaboration-potential

fully functional IoT. This is inline with the 3 pillars identified in the IoT Staff Working Document in order to advance IoT in Europe:

- A single market for the IoT: IoT devices and services (thus including data) must be able to connect seamlessly and on a plug-and-play basis anywhere in the European Union (EU), and scale up without hindering from national borders;
- A context of thriving IoT Ecosystems: new products and services in selected lead markets such as Industrial IoT, and the existence of open platforms across vertical silos, helping developers' communities to innovate and not causing lock-in situations for users;
- A human-centred IoT: European values must find their application for the IoT to empower citizens rather than machines and corporations, driven by high privacy and security standards and notably through a "Trusted IoT" label.

On the other side of the Atlantic, US Congress has introduced the Developing Innovation and Growing the IoT (DIGIT) Act to facilitate planning and coordination among government and private entities to support expanded use of the IoT.

The initiative considers that advances in technology could mean using the IoT to create life-improving developments for everything from health care to transportation to energy management to smart cities. The strategy aims to incentivise the development of the IoT, prioritise accelerating IoT's development and deployment and ensure it responsibly protects against misuse.

The DIGIT Act forms a working group consisting of businesses, non-governmental stakeholders, and federal agencies that would issue guidance on potential regulatory barriers, current and future spectrum needs, and possible security concerns. The resolution underscores the US's commitment to nurturing innovation, but also in protecting consumers and finding solutions to societal challenges through technology driven solutions like IoT.

The strong focus in both regions on implementing a strategy on IoT offers many opportunities for collaboration and cooperation in the years to come.

Additional international cooperation partnerships are expected with further partner countries or regions. In particular, cooperation with India is highly anticipated. India has created its vision "to develop connected and smart IoT based system for our country's economy, society, environment and global needs" and is rolling out its IoT action plan. The Indian IoT policy comprises of five vertical pillars (Demonstration Centres, Capacity Building and Incubation, R&D and Innovation, Incentives and Engagements, Human

Resource Development) and 2 horizontal supports (Standards and Governance structure). International cooperation is anticipated in several areas of the IoT policy programme. For instance, in human Resources Development, it is called for bilateral cooperation programs between Indian premier institutes and institutes of other countries. Europeis expected to approach authorities and academic/research institutes in India to explore synergies and collaborations for global solutions but also local exploitation.

Bibliography

[1] Report on the implementation of the strategy for international cooperation in research and innovation; COM (2012) 497.

[2] Commission Staff Working Document "Roadmaps for international cooperation", accompanying the document "Report from the Commission to the European Parliament, the Council, the European Economic and Social Committee and the Committee of the Regions/Report on the implementation of the strategy for international cooperation in research and innovation"; COM (2014) 276.

[3] EU-China Joint White Paper on the Internet of Things, China Academy of Information and Communications Technology (CAICT) & European Commission – DG CONNECT, January 2016.

[4] Global Mobile Economy Report 2015, GSM Association, 2015.

[5] GE, Analysis of the economic contribution of the industrial Internet to the United States, 2013.

[6] The second term of Ten application cases of IoT in Wuxi, 2014.

[7] CAICT, National IoT survey data, 2014.

[8] Beijing municipal commission of economy and informatization, open published data, 2014.

[9] ABDI. Mapeamento da Cadeia Fornecedora de TIC e de seus Produtos e Serviços para Redes Elétricas Inteligentes (REI) – Sumário Executivo. 2014.

[10] IST-Africa. Report on ICT Initiatives and Research Capacity, v2, 31 January 2016.

[11] Fiona Graham, "Thwarting Senegal's cattle rustlers using mobile phones". http://www.bbc.com/news/business-28365741

[12] Daral Technologie by Amadou Sow. http://www.agritools.org/stories/daral

[13] Marco Zennaro and Antoine Bagula, "IoT for Development (IoT4D)". IoT Newsletter, July 14, 2015. http://iot.ieee.org/newsletter/july-2015/iot-for-development-iot4d.html

[14] Wienke Giezeman, "Building a Crowdsourced Global IoT Network Operator". IoT Newsletter, January 12th, 2016. http://iot.ieee.org/news letter/january-2016/building-a-crowdsourced-global-iot-network-operat or.html

[15] William Toll, "Top 49 Tools for the Internet of Things". https://blog.profit bricks.com/top-49-tools-internet-of-things

[16] Opinion 8/2014 on the on Recent Developments on the Internet of Things, Article 29 Working Party on Data Protection, 14/EN, WP 223, 16 September 2014, online at http://ec.europa.eu/justice/data-pro tection/ article-29/documentation/opinion-recommendation/files/2014/w p223_en.pdf

[17] Internet of Things – Privacy & Security in a Connected World, FTC Staff Report, January 2015, online at https://www.ftc.gov/system/files/docume nts/reports/federal-trade-commission-staff-report-november-2013-work shop-entitled-internet-things-privacy/150127iotrpt.pdf

[18] UNIFY-IoT, online at www.unify-iot.eu

[19] AIOTI, online at www.aioti.eu

[20] PICASSO, online at www.picasso-project.eu

[21] IoT European Platforms Initiative, IoT-EPI, online at www.iot-epi.eu

[22] CAICT, Calculation data for MIIT of PRC, 2015.

Index

5G 9, 86, 111, 323

A
AAL 44, 117
Accountability 230, 232
Adaptive Gateway 266
Africa 295, 308, 312
Ageing Well 42
Agile 109, 114, 139, 170
Agile Food Factories 141
Agriculture 79, 132, 309
Agri-food logistics 129
AI 296
AIOTI 262, 322
Alliance for the Internet
    of Things Innovation 321
Ambient assisted living 44, 117
Ambient Intelligence 26
Ambient Localization 26
Ambient sensing 26
Analytics 34, 155, 180, 242
API 269, 280
Architecture 252, 261
Artificial Intelligence 23, 100
Automation pyramid 170, 174
Autonomic computing 30
Autonomous 24, 140
Autonomous Farm 140, 148
Autonomous vehicles 62, 11
Autonomy 26
Availability 42, 172

B
Barcodes 135, 282
Be-IoT 266, 288
Big Data 9, 75, 312
BigIoT 269, 298

BIoTope 270, 275
Blue economy 11
Brazil 9, 293, 304
Business 290, 301, 319
Business ecosystem 5
Business models 11, 78, 146, 290

C
China 117, 298
Circular Economy 11, 140
Citizen Centric 72
Cloud 41, 63, 81
Cloud Computing 30, 81
Cognitive 24, 107
Cognitive IoT 23, 24
Communication 12, 23, 57
Connected Supply Chain 154, 155
Connectivity 84, 140
Consumer 16, 39, 53
Consumer awareness 56, 132
Corrective Maintenance 175, 176
CPS 161, 171, 300
Cyber Physical 133, 185, 290
Cyber Physical
    Production Systems 157, 161
Cyber physical systems 37, 58, 111

D
Data 9, 23, 42, 73, 106
Data access 43, 227
Data ownership 55, 144, 227
Data protection 202, 224, 261, 325
Data security 57, 143, 325
Data-Driven Farming 79, 140
Deep learning 24, 109
Definition 23, 76, 173
Demand-driven Farming 140, 148

335

Lightning Source UK Ltd.
Milton Keynes UK
UKOW06n1147270117
293027UK00002B/38/P